SCANNING PROBE MICROSCOPY

SCANNING PROBE MICROSCOPY

Editors

Nikodem Tomczak
Kuan Eng Johnson Goh
*A*STAR, Singapore*

World Scientific

NEW JERSEY • LONDON • SINGAPORE • BEIJING • SHANGHAI • HONG KONG • TAIPEI • CHENNAI

Published by

World Scientific Publishing Co. Pte. Ltd.
5 Toh Tuck Link, Singapore 596224
USA office: 27 Warren Street, Suite 401-402, Hackensack, NJ 07601
UK office: 57 Shelton Street, Covent Garden, London WC2H 9HE

British Library Cataloguing-in-Publication Data
A catalogue record for this book is available from the British Library.

SCANNING PROBE MICROSCOPY

ISBN-13 978-981-4324-76-2
ISBN-10 981-4324-76-0

Typeset by Stallion Press
Email: enquiries@stallionpress.com

Printed in Singapore.

Contents

Biographies of Volume Editors

Nikodem Tomczak studied chemistry and polymer chemistry & physics in Gdańsk (Poland). He completed his PhD in Chemistry at the University of Twente (The Netherlands) with Professor Julius Vancso and Professor Niek van Hulst. His thesis topic was on the application of single molecule fluorescence methods to studies of polymer dynamics in thin films. During his postdoctoral period he was involved in the measurement of the nanomechanical properties of single molecules. He is currently a Senior Research Engineer at the Institute of Materials Research and Engineering (A*STAR) in Singapore. His major research interests include polymer chemistry and physics at surfaces and interfaces, single molecule fluorescence detection, semiconductor nanocrystals (Quantum Dots), and Scanning Probe Microscopy of polymers.

Kuan Eng Johnson Goh is currently a Senior Research Engineer at the Institute of Materials Research and Engineering (Singapore) where he leads the efforts in using scanning probe microscopy for atomic-scale fabrication and material interface characterisation. He completed his PhD in Physics at the University of New South Wales in 2006 under an Endeavour International Postgraduate Research Scholarship awarded by the Australian government and three International Fellowships awarded by A*STAR (Singapore). His thesis dealt with the encapsulation of Si:P delta-doped devices fabricated by scanning tunnelling microscopy and the magnetotransport in such devices. His current research addresses key issues in the fabrication of organic photovoltaic devices using ballistic electron emission microscopy, the reliability of high-K dieletrics, and the possibility of atomic precision manufacturing.

Preface

Nanotechnology is a broad topic of intense interest for fundamental and application-driven research. The Singapore R&D community has embraced nanotechnology and launched significant initiatives within the two major Universities (the National University of Singapore and Nanyang Technological University) as well as the local research institutes (established under the Agency for Science, Technology and Research). Put simply, the last decade has seen a large growth in resources and capability within Singapore for nanotechnology, and scanning probe microscopy (SPM) — the subject of this book — is part of that effort. The aim of the book is to provide a snapshot of recent SPM research both within Singapore and in collaboration with our international colleagues. Two broad themes, lithography and atomic force microscopy (AFM) in liquids, are presented.

Lithography at the atomic scale can use either SPM to manipulate single atoms and molecules into a desired structure, or molecular structures can be formed directly by self-assembly on a surface. Both approaches are described, along with ballistic electron emission microscopy (BEEM) — an important tool for understanding how to electrically connect metal wires to organic molecules. It is also essential, for realistic applications, to join the molecular scale to the micrometer scale, and various nanolithography methods based on SPM are discussed (e.g. dip pen lithography, localized heating, oxidation or electrochemical reactions).

AFM in liquid environments not only provides the ability to image surfaces in their natural state but equally important one can undertake force spectroscopy over highly localized regions. The term "force spectroscopy" is very general and refers to the measurement of any localized mechanical property, such as elasticity, dissipation or stiffness. Again, experiments over a broad length scale are presented in this book, demonstrating the utility of AFM force spectroscopy from single polymer chains and molecularly ordered liquids, to probing disease in red blood cells.

Finally, although the surface under study is often very well characterized, the tip condition is typically unknown. This remains a general and major problem in SPM. At the start of this review volume, Chapter 1 is appropriately devoted to efforts at producing well-defined tips, which range in length scale from single atom tips for molecular lithography to macroscopic spherical tips for force measurements.

I would like to thank the Editors for their considerable effort in bringing together some of the diverse aspects of SPM research in Singapore and I am sure the reader will find the topics chosen rewarding and stimulating.

Sean O'Shea
IMRE, Singapore
15 March 2010

Nanotip Technology for Scanning Probe Microscopy

Chapter 1

Moh'd Rezeq* and Christian Joachim†

Abstract

Nanotips have been in increasing demand over the last two decades, owing to their crucial role in the main nanotechnology instruments, like scanning probe microscopes and scanning electron and transmission microscopes. Well-defined ultrasharp tips have also attracted a growing interest in molecular electronics, as they are needed for fabrication and characterization of molecular structures. In this chapter we present several methods that have been explored so far for fabricating such ultrasharp tips, with an emphasis on the most recent technique. We also highlight some of the applications of these tips in various aspects of nanotechnology.

1.1. Introduction

The invention of scanning probe microscopes, i.e. the scanning tunneling microscope (STM) and the atomic force microscope (AFM), has raised a continual need to develop well-controlled methods for fabricating ultrasharp tips with a radius of a few nanometers, which are often referred to as nanotips. Before we describe the fabrication and characterization of

*Institute of Materials Research and Engineering (IMRE), 3 Research Link, Singapore 117602, and Khalifa University of Science, Technology and Research (KUSTAR), UAE, Email: rezeqm@gmail.com

†Institute of Materials Research and Engineering (IMRE), 3 Research Link, Singapore 117602 and CEMES-CNRS, 29 rue J. Marvig, PO Box 4347, F-31055 Toulouse Cedex, France.

nanotips, it is worth introducing the reader to the early inventions of remarkable and simple ultrahigh resolution microscopes that allow the characterization of sharp tips.

The first technology is the field emission microscope (FEM). The phenomenon of field emission was first discovered in 1897. Its rigorous theory was developed in 1928 by Fowler and Nordheim based on quantum-mechanical tunneling of electrons through a surface barrier exposed to a high electric field.[1] Muller[2] was able to obtain such a high electric field by applying a few kV on a nominally 100-nm-radius metal tip, which led to the invention of the field emission microscope (FEM) (also called the field electron microscope) in 1936, with a resolution of a few nanometers. In 1955 Muller and his PhD student Bahadour were able to demonstrate the first atomic resolution microscopy by using a reversed tip voltage polarity in the presence of an inert gas (like He) and maintaining the tip at liquid nitrogen temperature.[2,3] A new type of microscopy emerged, which was later referred to as the field ion microscope (FIM).[2] The use of these microscopes is still limited to the characterization of sharp objects, typically less than 0.1 μm, as a high electric field (in the range of 5 V/nm and 50 V/nm for the FEM and the FIM, respectively) is required for the imaging process. These microscopes have also been used to image atoms, molecules and small structures deposited from an external source or diffused from the shank onto the tip apex surface.[4–23] Muller also proposed, in 1967, a remarkable application of the FIM for the analysis of atomic species and called it the atom probe microscope. The design of this microscope is based on a combination of an FIM and a time-of-flight mass spectrometer, which is placed behind the FIM screen. A small hole is located at the center of the screen for ions to pass through into the mass spectrometer. Atoms, or probably small molecules, will in turn be selected by the FIM and then field-evaporated and sent into the mass spectrometer for identification. Furthermore, the three-dimensional tip structure can be reconstructed using a computer program by evaporating (and hence destroying) the tip layer by layer.[24–27]

There has been renewed interest in the FIM since the advent of the STM and the AFM in the early 1980s. Both the STM and the AFM allow atomic scale resolution on flat surfaces. Well-defined sharp tips are demanded for ultimate resolution in the SPM, as well as for manipulating atoms, molecules and nanostructures, paving the way for the era of nanotechnology.[28–45] However, the progression of nanomaterial characterization, nanoscale device fabrication and other aspects of nanotechnology relies on finding a reliable technology for nanotip fabrication. The capability of the FIM to clean and characterize tips with atomic resolution motivated researchers to go beyond the capability to shape and sharpen this tip apex. Indeed, they have succeeded in developing several methods over the last two decades for fabricating ultrasharp tips that terminate with an apex radius in the order of 1 nm and sometimes with a single atom apex.[46–53] We will see in the next sections that the fabrication of such well-defined ultrasharp tips can be rigorously designed and engineered in the FIM. Nanotips are also advantageous in the transmission electron microscope. Here, atomic resolution requires a highly collimated and bright electron beam and these two parameters are direct functions of the tip size.[54,55]

In this endeavor, we will summarize the methods that have been developed so far pertaining to the fabrication of nanotips with a greater focus on the most recent one by Rezeq *et al.*,[52a,b] which is based on a spatially controlled field-assisted reaction. We find that it is more instructive to start with a brief introduction of the principles of field electron and ion microscopy, which are detailed in several publications.[1,2,24,25] Then we introduce the reader to the well-known methods of nanotip fabrication, which are:

(i) The field–surface melting and build-up method;
(ii) The ion sputtering and deposition method;
(iii) The Pd-covered single atom tip;
(iv) Field-enhanced diffusion growth;
(v) The spatially controlled field-assisted etching technique.

All of these methods demonstrate the fabrication of single atom tip ends, as directly characterized by the FIM. These single atom tips also have an identical field emission pattern represented by a confined FEM spot on the screen. Besides the numerous applications of these ultrasharp tips in SPM machines, we have to mention that these tips are also ideal electron sources for lensless projection and holographic microscopes as they produce quite coherent and bright electron beams at low energies.[46,56,57]

In fact, atomic and molecular manipulations on flat metal or semiconductor surfaces with the STM often result in a change of the atomic structure of the tip apex. This undesired consequence raises the need for a reliable technique that permits the restoration of the tip end structure. We demonstrate here that after losing its last single atom, the tip apex can be modified atom by atom to restore the single atom tip by using the spatially controlled field-assisted etching method.

One of the merits of the field-assisted etching method is that it applies to any crystal orientation or polycrystalline metals as well,[58] unlike the other methods that are preferentially used on W(111) crystals.

1.2. Field Electron Microscope (FEM) and Tip Characterization

Field electron emission is a quantum phenomenon where an electron tunnels from the Fermi level of a metal surface into the vacuum when a high-enough electric field (nominally a few V/nm) exists at the metal–vacuum interface. Therefore, this kind of microscope is often called a field emission microscope. In contrast, an amount of energy, usually referred to as the work function (ϕ), is required to eject an electron from the Fermi level into the vacuum in the absence of any electric field, as illustrated in Fig. 1.

Normally, the electrical potential distribution in the vicinity of a metal surface results from the image force and is given by $U(x) = -e^2/4x$, where e is the electron charge and x is the distance from the surface. When a high electric field (F) is applied, the barrier height is lowered and the barrier width becomes narrow enough for electrons to tunnel. The potential function is then $U(x) = -e^2/4x - eFx$. Such high fields can be attained by applying a negative voltage of −0.2 to −2 kV on a sharp tip with a radius of less than 100 nm (some researchers prefer to use an alternative arrangement where a positively biased extractor is placed close to a grounded tip at probably a lower voltage). This mechanism is referred to as "cold field emission," assuming that there are no thermally excited electrons in higher states as in the case of Schottky emission, and is governed by the Fowler–Nordheim (FN) theory.[2] The field emission microscope comprises a phosphor screen that is usually supported by a set of microchannel plates

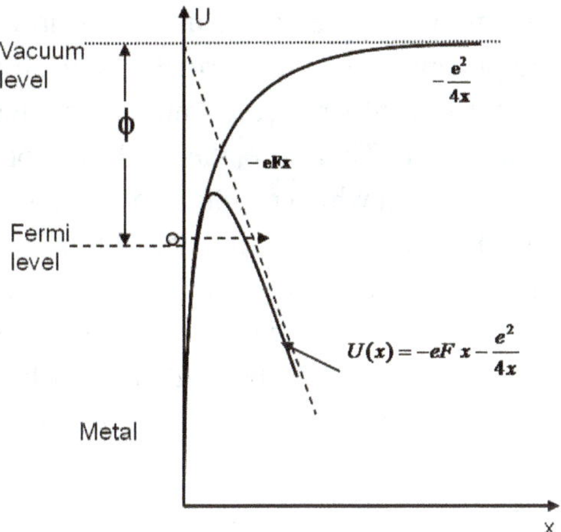

Fig. 1. Illustration of the potential energy distribution near a metal surface. At a strong electric field F the barrier becomes narrow enough for an electron to tunnel into the vacuum.

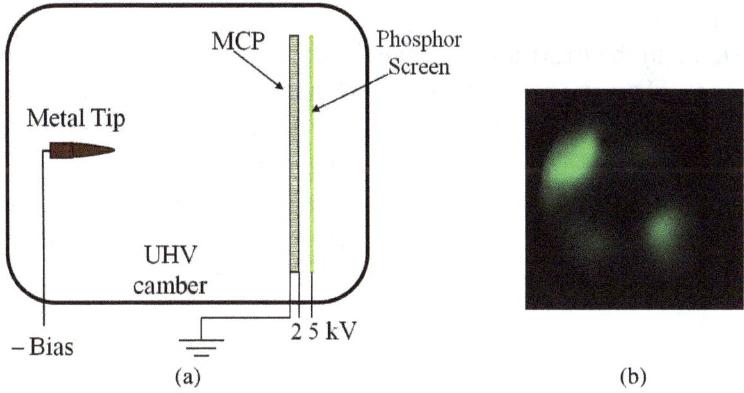

Fig. 2. (a) Schematic of the field emission microscope. (b) FEM image of a bcc W tip; the dark areas are from the flat {110} planes, and the bright areas from {111}, {121}, {310} and other high index planes.[3]

(MCPs) as an image intensifier [Fig. 2(a)], and a tip placed 2–4 cm from the screen, and the entire setup is mounted in an ultrahigh vacuum (UHV) chamber. The MCPs, the phosphor screen and the tip are connected to the external high voltage power supplies via UHV-compatible electrical feedthroughs.

The electrons emitted from the tip end usually follow the electric field trajectory and hit the phosphor screen. Normally, a clean tip with a hemispherical shape is composed of several atomic planes that have different local work functions and radii. Therefore, the field electron emission intensity has a local spatial variation which results in bright and dark areas in the field emission pattern. For instance, on tips with a bcc lattice structure the closely packed low index (110) plane has a higher work function than (111), (121) and other higher index planes and thus looks darker than other planes; see Fig. 2(b). These atomic facets can be readily identified in the FIM, as will be discussed later. The magnification (M) of the FEM is given by

$$M = \frac{1}{\gamma} \cdot \frac{L}{R}, \tag{1}$$

where L is the distance between the screen and the tip, R the tip radius, and γ a compression coefficient (1.5–2). It is used to describe the deviation of the electron trajectory from the normal radial path, which in turn depends on the tip-screen geometry.[1]

Although the FEM image provides information about the tip surface roughness and crystal orientation, the FN theory [1,2,59–62] can also be used to estimate the tip radius and average work function. The derivation of the FN theory is detailed in several publications.[1,59] Here, we only describe briefly how this theory is used to extract some characteristic tip parameters and the other theories that have been proposed to avoid the limitations of the FN theory. The general form of the FN equation can be simply written as[2,59]

$$\frac{I}{V^2} = \frac{a}{R^2} \exp\left[-\frac{bR}{V} v \left(\frac{cV^{1/2}}{R^{1/2}}\right)\right], \tag{2}$$

where V is the applied tip voltage, R is the tip radius, $a = 6.16 \times 10^{-8} A$ (A is the emission area), $b = -3.415 \times 10^8 \phi^{3/2}$, $c = 1.8 \times 10^{-4}/\phi$, and the function v can also be approximated to 1 when the first order term of the Taylor expansion is assumed.[2,63] It is obvious from the above equation that the semilog plot of (I/V^2) vs. $1/V$ should be linear. Either the tip radius or the work function can be calculated from the slope of the line assuming that one of them is known in advance, whereas the emission area may be extracted from the data line intercept.

It is worth stating that Eq. (2) was primarily derived for planar emitters, and thus $\ln(1/V^2)$ vs. $1/V$ is linear, and this formula can be rather conveniently applied to broad tips (radius >20 nm). For sharp tips (radius <20 nm), the I–V data do not fit the standard FN theory.[1,59,60] This deviation is attributed to the assumptions that have been considered in the derivation of this equation: (1) the field is uniform in a plannar geometry, (2) the work function is uniform over the emitting area, (3) a one-dimensional barrier is considered, (4) the calculations assume that temperature $T = 0$, and (5) the potential barrier width is much less than the emitter radius. However, modified forms of the FN theory were introduced for tips of radius <20 nm. Culter *et al.*[59] found that I–V data of a hyperboloidal tip obey the formula $J = A\,V^2\exp(-B/V - C/V^2)$, where A, B and C are constants which depend on the material and geometry of the tip. The calculations by Yuasa *et al.*[60] for a similar geometry led to a different formula and they found that $\frac{I}{V^3} \propto \exp(\frac{-A}{V})$, where A is a constant that depends on the tip–collector distance and tip sharpness. These modified formulae provide a better I–V data fit for a tip radius of 10–20 nm. But they are still not accurate for a tip radius of 1–10 nm, due to the enhancement of field emission.

In conclusion, the standard FN theory applies adequately to tips >20 nm, whereas the modified expressions have to be used for sharper tips down to 10 nm. For extremely sharp tips more accurate methods for estimating the tip radius are still required. These methods will be addressed in Sec. 1.3 and Subsec. 1.7.2.

1.3. Field Ion Microscopy (FIM)

The principles of field ion microscopy are essentially different from that of field electron emission. In the latter, an inert gas is introduced into the system for imaging while the tip is held at a high positive voltage and cooled down to the liquid nitrogen temperature. To convert the FEM to an FIM, more components are added, like a cryostat for tip cooling and a precision leak valve for introducing inert gases into the UHV chamber for imaging. Noble gases like helium, argon and neon, and diatomic gases like hydrogen, can be used as imaging gases as well. The setup is similar to the one depicted in Fig. 2.

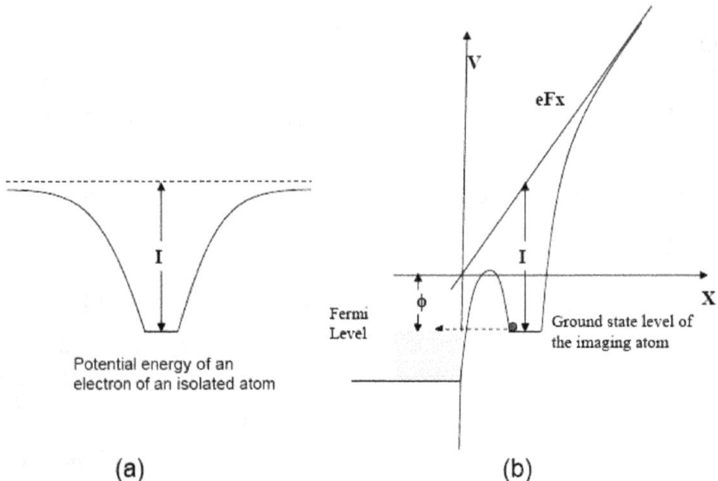

Fig. 3. (a) The potential well of an isolated atom with ionization energy *I*. (b) The existence of the atom in the vicinity of a metal surface that is exposed to a strong electric field (F) results in a very narrow potential barrier and hence a probability of an electron tunneling from the ground state of the atom into the top of the Fermi level. ϕ is the work function of the metal.

It is worth describing the ionization mechanism of a gas atom in the presence of a strong electric field on the metal surface. The potential well of an isolated atom is illustrated in Fig. 3(a), where I is the ionization energy. When this atom is brought to the vicinity of a metal surface, the atom–metal barrier becomes much narrower and is reduced further in the presence of a strong electric field. The tunneling probability will be significantly enhanced when the potential barrier becomes comparable with the electron wavelength associated with the atom [Fig. 3(b)]. The tunneling probability in this case is represented by the Wentzel–Kramer–Brillouin (WKB) theory[2]:

$$P(E, V(x)) = \exp\left\{-\left(\frac{8m}{\hbar^2}\right)^{1/2} \int_{x_1}^{x_2} [V(x) - E]^{1/2} dx\right\}, \qquad (3)$$

where m is the electron's mass, $V(x)$ and E are its potential and kinetic energy respectively, $V(x)$ is a function of position x, $\hbar$ is Planck's constant divided by 2π, and x_1 and x_2 are the edges of the barrier at energy level E. The tunneling of an electron from the imaged atom strictly occurs at a critical distance x_c from the surface,[2] which is given by

$$x_C = \frac{I - \phi}{eF}. \tag{4}$$

In the case of a tungsten tip ($\phi = 4.5\,\text{eV}$) and where He is used as an imaging gas, with $I = 24.5\,\text{eV}$ and F (the ionization field) $= 4.4\ V/\text{Å}$, the typical value of x_c is 4.5 Å.

Primarily, the gas atom in the vicinity of the tip surface gets polarized due to the strong electric field strength and is consequently attracted to the tip and will reach the surface at a high kinetic energy. The accommodation time of the attracted atom on the apex surface is not adequate for electrons to tunnel into the metal. Therefore, this polarized atom hops on the surface several times, losing some of its kinetic energy until it becomes slow enough for the tunneling process to occur, as illustrated schematically in Fig. 4(a). Once the atom becomes a positive ion, it gets expelled from the ionization point (i.e. the surface atom) toward the screen (a combination of MCPs and a phosphor screen), creating a bright spot. These events take place all over the tip surface. The expelled ions follow the trajectory of the electric field lines and create a two-dimensional (stereographic) projection image on the screen with a spectacular feature of concentric rings of atomic resolution. Each set of concentric rings actually represents a particular lattice facet. Cooling the tip down to the liquid nitrogen temperature usually enhances

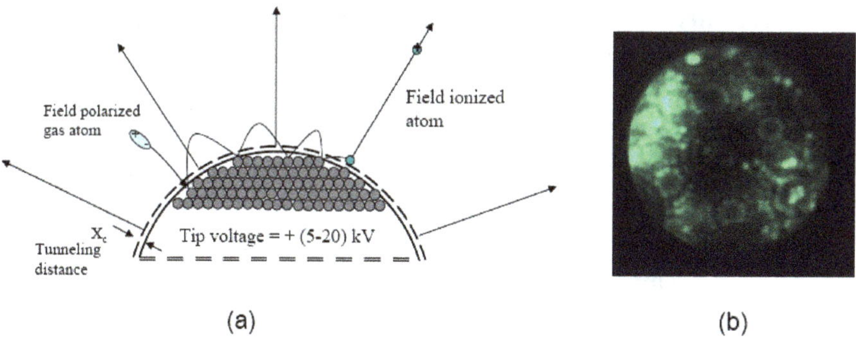

Fig. 4. (a) Schematic illustration of the FIM imaging process. The atom gets polarized and then hops on the surface before getting ionized by an electron tunneling into the metal at a distance x_C. (b) An FIM image of a bcc single crystal tungsten tip. Each single bright spot on the image is an atom. The flat plane at the center (110) represents the orientation of the crystal.

the ionization process, as gas atoms get localized faster and the surface tip atoms are more stable, which results in a substantial improvement in the FIM resolution and contrast. On tips of a bcc lattice structure, the closely packed facets with low index parameters, like (110), show wider atomic planes than that of the low index facets, like (111); see for example Fig. 4(b).

The electric field strength required to ionize the imaging gas atom is denoted as the best imaging field (F_0) and is be given by

$$F_0 = \frac{V_0}{\kappa R}, \tag{5}$$

where V_0 is the best imaging voltage, R the tip radius and κ a characteristic factor $\approx$ 4–6.[2] The FIM resolution is also proportional to $1/R$, and hence the sharper the tip the higher the resolution that can be obtained.[2,24] The resolution of the FIM has been rigorously calculated for different imaging gases at different temperatures. A resolution better than 2.5 Å, i.e. atomic resolution, has been reported, when helium was used as an imaging gas for a tip radius of around 10 nm.[2,24,25]

1.4. Preparation and Characterization of an Atomically Clean Tip in an FIM

Metal tips initially prepared using the common electrochemical etching method very likely exhibit multiple miniprotrusions, which can be seen in the FIM. This tip roughness limits the performance of these tips in either TEM or SPM systems. Therefore, obtaining an atomically clean tip apex would be an advantage. Before the tip can be used in any UHV microscopy system, various contaminations must be carefully cleaned off it. This cleaning (or annealing) process can be done in two different ways.

The first method is called electron bombardment (or e-beam heating). Here, a positively biased tip is placed in front of an extremely hot tungsten filament (filament temperature = 2000–2500°C). The electrons coming out of the filament are attracted by the positive sharp tip end, due to the strong electric field strength. This results in heating the tip apex to a substantially high temperature and therefore clearing the tip of oxide and other contaminating layers.

The second method is the direct tip heating process, where the tip is spot-welded on a triangular or curved wire with two ends connected to a power source by UHV electrical feedthroughs. The supporting wire is resistively heated and it in turn heats the tip to the annealing temperature.

The above two methods are good for cleaning off contaminations from the tip but not effective for removing surface defects or corrugations. We show here that the FIM is effective for removing all undesired protrusions via the "field evaporation process". When the electric field is increased beyond the best imaging value, the metal atoms of the tip start to get ionized and subsequently evaporate from the tip surface; this process is referred to as "field evaporation." Fortunately, the enhancement of the electric field on the local sharp protrusions and edges makes it possible to strictly remove (field-evaporate) these sites first, before the atoms from the tip's base. To ensure an atomically clean surface, the applied voltage should be gradually and carefully increased until all atomic protrusions and corrugations are removed. At this stage, the field evaporation stops through a balance between the tip curvature and the applied voltage at the best imaging field. For this demonstration, we used a treated tip made from a polycrystalline tungsten wire. We found that in a reasonable number of cases the field evaporation resulted in a single crystal end with (110) orientation, with the (110) facet appearing at the center of the FIM image, as in Fig. 4(b). In some cases an apex of multiple grains with visible boundaries can be seen.

As demonstrated in Fig. 4(b), the FIM image is usually composed of sets of concentric rings where each set is oriented at a particular pole (orientation). It is best represented by a stereographic projection of a sphere of the imaged lattice material, for instance (bcc) or (fcc) lattice materials.[2,24,25] Standard stereographic maps of (110) bcc and (001) fcc are available in some textbooks.[2,25] The size of a plane in the FIM is proportional to the surface atom density of that plane. Therefore, closely packed planes with higher atomic steps are more prominent and larger in FIM images than the ones with less atomic density and very likely have high index parameters. In the bcc structure, these planes form the sequence {110}, {100}, {211}, {310}, {111} . . . (from largest to smallest). In the stereographic projection map, the sizes of planes are presented according to this fact. Therefore, by comparing the relative change of the plane area on the

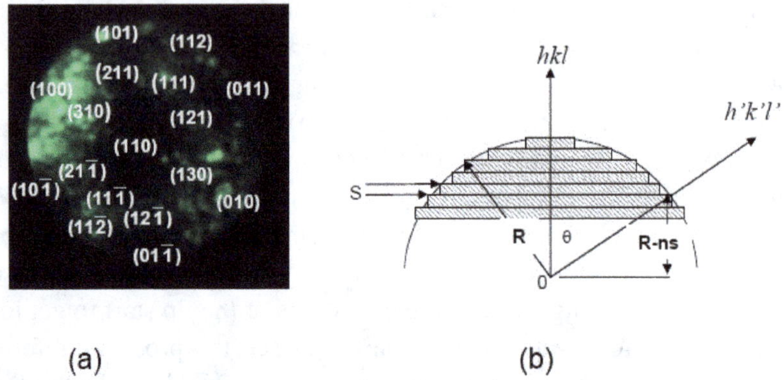

Fig. 5. (a) Indexed atomic facets labeled on the FIM image. (b) Schematic showing the atomic steps of the tip apex. *S* is the spacing between successive steps and *R* is the tip radius. θ is the angle between two consecutive poles (*hkl*) and (*h′k′l′*), and *n* is the number of atomic steps.

FIM image with that on the stereographic map, one can readily identify the crystal facets and thus the orientation of the tip. The main facets of the FIM image in Fig. 4(b) are labeled according to this approach and illustrated in Fig. 5(a).

In a given direction, the step height can be calculated from

$$s = \frac{a}{\delta\sqrt{h^2 + k^2 + l^2}}, \tag{6}$$

where (hkl) are Miller indices for a particular plane, s is the distance between two successive planes, a is the lattice constant, and $\delta = 1$ if $(h + k + l)$ is even and $\delta = 2$ if $(h + k + l)$ is odd for bcc structures. Thus, with a simple analysis, as illustrated in the schematic in Fig. 5(b), the tip radius can be calculated to a good approximation using the formula for s:

$$R = \frac{ns}{1 - \cos\theta}, \tag{7}$$

where θ is the angle between two poles (atomic planes) and n is the number of atomic steps. For tungsten tips, one can calculate the tip radius by counting the number of rings from the (110) plane to the center of the (121) plane and then multiplying by 16 Å. Accordingly, the tip radius in Fig. 5(a) is equal to ~96 Å.

1.5. Brief Review of Previous Nanotip Fabrication Methods

1.5.1. *Field–surface melting method and build-up method*

The field–surface melting method is based on heating the tip to a temperature below the bulk melting temperature $T_m(\sim\frac{1}{3}T_m)$ in the presence of a strong electric field. At this elevated temperature the surface atoms are highly mobile under the influence of the electric field, which reduces the activation barrier for diffusion of these atoms. When any protrusion or roughness (a few nm high) exists on the tip surface, a field gradient is formed on these sites. This field gradient can be large enough to drive the mobile atoms over that protrusion to form a much sharper end and can terminate with a single atom.[46,63] If the surface is quite flat the effect of the surface diffusion will be negligible even at a strong electric field. Fortunately, the variation of the local radii of various planes on a clean tip surface results in a higher electric field on the high index facets. This forms the basis of the "build-up" method.[47] The surface diffusion of atoms of some low index planes along the field gradient leads to the enlargement of these planes. For the tungsten (111) single crystal orientation, the build-up method leads to the enlargement of the {121} planes and stops when two neighboring planes meet, forming a pyramid-like shape. A further increase of the electric field leads to the ionization of the topmost atom and consequently evaporates it from the top end. A well-confined ion beam from the metal atoms can be produced by careful temperature and field adjustments in a process referred to as atomic metallic ion emission (AMIE).[64] The pseudostationary profiley (PSP) technique[47] is very much like the AMIE method but, in addition to achieving a steady state between the surface diffusion of atoms to the tip apex and the field evaporation of atoms from the tip end, this process takes place in the presence of oxygen and at a high temperature. Hence, the formation of a single atom tip cannot be verified in an FIM until the tip is cooled down. Often, a tip prepared this way exhibits a three-atom apex.[47] Moreover, as this method is based on the diffusion of atoms from the shank to the top of the apex, other contaminations will likely be involved.

1.5.2. *Deposition of an external metal atom on tips sharpened by ion sputtering*

In this method the tip is sharpened first by bombarding the apex with ions created and guided by the electric field on the tip. This process is based on applying a negative voltage to the tip and introducing a certain amount of an inert gas like neon into the UHV chamber. The electrons emitted from the tip ionize the gas atoms and thus create positive ions, which are subsequently attracted to the negative tip. The collision of ions with the metal surface at a high kinetic energy leads to the removal of metal tip atoms, and hence sharpens the tip. Maintaining a constant field emission current requires a careful reduction of the applied voltage, due to the tip sharpening. The continuous sputtering process can lead to a minimum tip size, in the order of a few nanometers. This sharp tip is subsequently annealed to remove any adsorbates. Field evaporation of a few atomic layers may be required to improve the tip surface. Fink *et al.*[48] were able to control the field evaporation of a previously ion-sputtered W(111) tip to obtain a trimer apex. Later, a single W atom from an external metal source was thermally deposited on the trimer, forming a single atom end.[48] However, the process of strictly depositing one atom at the center of the trimer might not be sufficiently controlled. In addition, other atoms are expected to settle down in the vicinity of the atom at the center of the apex.

1.5.3. *Pd-coated tungsten single atom apex*

In this method, a clean W(111) tip surface is first coated with a few monolayers of Pd using thermal deposition. Then the tip is annealed at 1000 K for around 30 min, which leads to the formation of a pyramidal tip shape with single atom wedges where {112} facets meet with a top end along the ⟨111⟩ direction. This method has been developed by Fu, Tsong and other researchers.[49,50] The preferable {112} facet formation of the deposited metal film is attributed to the enhanced anisotropy of the surface free energy, as the base tip orientation is (111).[50] This technique can be applied to other metal thin films, like Pt, Au and Ir. The bottom (111) structure of the tip can be verified by field-evaporating the topmost single atom end, which should reveal the usual trimer-like structure. Further field evaporation reveals larger atomic planes along with the atomic wedges of the pyramidal tip, as reported in Ref. 49. This method has been developed further to enhance the reproducibility by coating the tip with a thin layer of Pd out of vacuum using an electroplating technique and then transferring the tip into the vacuum system and performing the standard annealing process.[50] The tip is reported to be stable up to a temperature of 1000 K. It can also be regenerated in vacuum by a similar annealing procedure when the single atom at the end is lost. However, the formation of the single atom tip cannot be confirmed unless the tip is characterized again at atomic level in the FIM.

1.5.4. *Field-enhanced diffusion growth technique*

This straightforward method is based on common observations in field ion and emission microscopy where adsorbates on the tip shank diffuse along the field gradient toward the tip end. However, the experiment has to start with an atomically clean tip surface characterized in the FIM. Then, the tip is left in a base vacuum around 1×10^{-7} Pa for one night. Subsequently, an increased negative bias is applied to the tip, which leading to the diffusion of adsorbed residual gases from the tip shank onto the apex. This can be recognized from the progressive change in the FEM pattern that converges from a blurred one to a single bright spot.[51] The preferred crystal structure is (111) tungsten. As the nanotip is formed from residual gases, neither the tip apex structure nor the material is known.

1.6. Mechanisms of Nitrogen Adsorption on Metal Surfaces

Researchers have been studying the adsorption of nitrogen on metal surfaces particularly tungsten ones) for several decades,[65–77] using various surface characterization methods, like thermal desorption, work function measurements by contact potential and Kelvin probe, low energy electron diffraction (LEED), electron energy loss of fine structure (EELFS), Auger electron spectroscopy (AES), X-ray photoelectron spectroscopy (XPS), field emission microscopy (FEM) and field ion microscopy (FIM). They have been able to classify the adsorption of nitrogen on W into three major categories:[65,66,68–73]

(i) The strongly bound β nitrogen state, which results from the chemisorption of nitrogen atoms desorbing as N_2 above 1000 K;
(ii) α nitrogen, which is originating from the chemisorbed N_2 and desorbing at 300–400 K;
(iii) γ nitrogen, when nitrogen is adsorbed as nitrogen molecules and desorbing below 200 K.

There is some agreement between these methods based on the adsorption of nitrogen on various atomic facets but disagreement abounds regarding the adsorption of N_2 on the most closely packed W(110) plane.

Madey *et al.*[66] used thermal desorption and XPS techniques to study the adsorption states of nitrogen on polycrystalline W. They reported that nitrogen adsorbed rapidly on polycrystalline W at 300 K, forming two strongly bound states, β_1 and β_2, which desorb at ~1000 K. When W is cooled to ~100 K, a weakly bound molecular γ state can be formed. Their XPS measurements also showed that a considerable amount of the β state exists even with adsorbed nitrogen at 100 K on polycrystalline W. In addition, they found that a part of the weakly bound γ-N_2 desorbs upon heating the substrate to 300 K and the other portion is converted to a β state.[66] Scientists presume that γ nitrogen and β nitrogen may occupy similar sites. Their conclusion is based on a comparison between β nitrogen desorption experiments following adsorption at 115 K, where γ-N_2 is stable, and repeating the measurements at 300 K, where γ-N_2 is not adsorbed.[65] Madey and Yates[72,73] used isotopic mixing and flash desorption techniques on polycrystalline W, to confirm the split of the β state into two substates, β_1 and β_2, as observed by XPS. The desorption of the β_1 state is found to obey first order kinetics, while the more tightly bound β_2 state desorbs with higher order kinetics. The β_2 state forms normally at lower coverage, whereas the more weakly bound β_1 state is formed at increasing exposure to nitrogen.

Ota and Usami[67] used EELFS to study the nitrogen chemisorption on W(100) and to obtain information about the interatomic distance between

W atoms and nitrogen. After taking measurements spaning a wide range of nitrogen exposure (1–210 L) to a W(100) surface, they found that the nitrogen–tungsten distance remained almost constant, about 0.145 nm, at early stages of the adsorption to 3 L, and increase significantly after that. They also found that the work function of the (100) plane decreases tangibly at the initial stages of the adsorption, <5 L, and then increases with increase of the absorbed nitrogen. This led to the conclusion that at low doses nitrogen diffuses into the top atomic layers of tungsten, which results in weakening of the surface structure and hence lowering of the work function. On the other hand, increasing the adsorption further will increase the work function as nitrogen diffusion into tungsten will stop, as confirmed by their Auger data for the nitrogen-to-tungsten ratio, which shows saturation above 5 L of nitrogen exposure.[67] Delchar and Ehrlich[69] have performed contact potential measurements on single crystal W surfaces exposed to doses of nitrogen at various temperatures. Their observations revealed a reduction in the work function of W(100) by 0.4 eV at low coverage, indicating a strong binding β state, which is in agreement with the observation by Ota and Usami,[67] in contrast to an increase in the work function by 0.15 eV reported for W(111), possibly due to α state contribution. Serrano and Darling[68] have computed the potential energy surface for the dissociation reaction path of nitrogen molecules on W(100). They demonstrated that the most favorable chemisorption sites of these molecules are the fourfold hollow sites. They also found that the dissociation of the molecule to occupy two hollow sites is exothermic by $\sim$3.2 eV, whereas the dissociation to atoms chemisorbed on atop sites is endothermic by $\sim$0.7 eV.

Delchar and Ehrlich[69] have observed no change in the work function of W(110) at room temperature, and hence no chemisorption of nitrogen on W(110). They found that a special case of γ nitrogen adsorbed at temperature $T \sim 110$ K in an atomic state and resulted in a lower work function of (110) by 0.15 eV. This agrees with observations by Yates *et al.*[65] of a molecular formation of γ nitrogen with a decrease in the work function of $\sim$0.19 eV on the closely packed W(110) plane. In contrast, the results by Tamm and Schmidt,[70] who were using flash desorption mass spectrometry, showed that β nitrogen is formed on W(110) at the initial stages of the adsorption with a saturation density of about 37% of that on W(100). They also found that the desorption activation energy E_d of nitrogen on W(110) is 79 kcal mol^{-1}, unlike the data reported by Delchar and Ehrlich, <9 kcal mol^{-1} for W(110) and 75 kcal mol^{-1} for W(100) and W(111).[69] α-N_2 has also been shown to be preferentially formed on W(111) and not on W(110) or W(100).[65]

When the tungsten cluster is small to a point where atomic planes are not recognized, two nitrogen adsorption states are identified: the weakly bound state is observed from clusters formed at liquid nitrogen temperature and the strongly bound state from clusters formed at room temperature.[71]

The above analyses do not consider the effect of a strong electric field applied on the metal surface on molecular adsorption and dissociation mechanisms. This is in fact the case when a high voltage is applied on a sharp tip in field emission and field ion microscopes. It was found that a strong electric field results in weakening of the binding energy and thus in dissociation of the molecule. The ionization field, and hence the evaporation field, of most molecules is usually less than the imaging field of the common imaging gases in the FIM. Consequently, the FIM is inadequate for molecular imaging.[2]

However, the field-induced dissociation of nitrogen was observed in the atomic probe microscope (APM).[76] Two different desorption field values (<2.5 V nm^{-1} and 48 V nm^{-1}) were obtained for nitrogen at 50 K, indicating two types of adsorption states: physically adsorbed and trapped states. It was also found that the dissociation of molecular ions requires a lower field, <10 V/nm.[75] In an FIM direct corrosive reaction of nitrogen adsorbed on metal surfaces (like tungsten, platinum and rhodium) was observed on low field areas where the molecule can penetrate the ionization barrier and then chemisorbs and dissociates.[74] This metal atom removal at relatively low electric fields occurs due to the weakness of the metal–metal bonds on the surface by the chemisorbed gas.

In light of these investigations, we can conclude that nitrogen adsorbs on all tungsten facets in different states. The strong dissociation state takes place on most planes with different rates, except for the (111) plane. This dissociation process is often accompanied by the displacement and protrusion of surface atoms, which is in turn enhanced with the presence of a strong electric field and consequently results in corroded surfaces.

1.7. Controlled Field-Assisted Etching Method for Tip Sharpening

The natural distribution of the electric field on a sharp tip when a high voltage (5–20 kV) is applied is always maximum on the apex and decreases substantially along the shank. This fact makes it possible to develop the controlled field-assisted etching method.[52] To obtain a good image for an atomically clean tip in an FIM, the tip apex is maintained at a field adequate for ionizing He atoms and denoted as the best imaging field (F_0); the typical value of this field is tabulated in several text books[2,25] ($F_0 = 4.4$ V/Å for He). As the desorption field, and hence the ionization field of nitrogen, is less than F_0, nitrogen molecules are unable to reach the apex and penetrate the ionization barrier, but instead they reach and settle on the tip shank, where the electric field is weaker, as depicted in Fig. 6(a). After the nitrogen molecule settles on the periphery of the apex, it dissociates and diffuses into the top atomic layers. This dissociation process is also assisted by the strong electric field, $F < F_0$, on the shank. The diffusion of nitrogen atoms causes the protrusion of the topmost W atoms out of the surface, which results in the enhancement of the electric field to a point adequate for field evaporation, as schematically illustrated in Fig. 6(b). The removal of the protruding W atoms from the shank leads to a sharper tip, which results in a stronger field at the apex. To maintain the apex field at F_0 whilst avoiding the removal of apex atoms, the applied voltage on the tip has to be reduced accordingly. A careful and continuous adjustment of the applied voltage while a certain nitrogen dose is maintained in the FIM chamber will confine the reaction to the tip shank and this will therefore lead to tip sharpening.

1.7.1. *Experimental setup and results*

A general description of the field ion and emission microscope setup has already been given in Secs. 2 and 3. For performing the N_2 etching process, this setup is equipped with some additional tools. Apart from the helium dosing line, another high precision leak valve is attached for the molecular nitrogen gas. Helium and nitrogen gases are cryopurified as they pass through the respective dosing lines. Various gas levels in the FIM

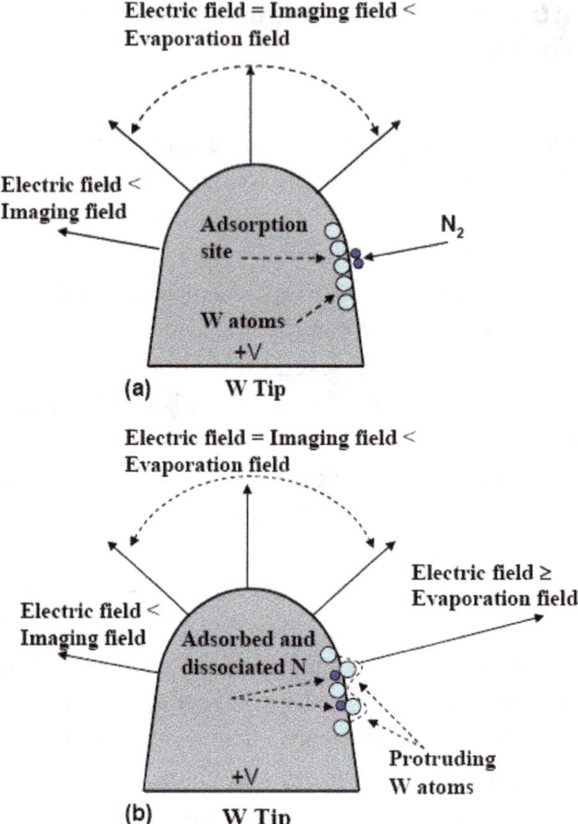

Fig. 6. The distribution of the electric field on the surface of a clean FIM tip at a fixed applied voltage when the best imaging field is attained on the apex. (a) The electric field along the shank descends below the threshold imaging value. The adsorption site on the shank, where the field is low, is indicated by a dashed arrow. (b) Illustration of atomic scale protrusions created as a consequence of the adsorption and dissociation of nitrogen, which in turn results in an enhanced electric field. This field is adequate for evaporating W atoms from this site.

chamber are continuously monitored by a quadrapole mass spectrometer to maintain a certain level of He and N_2 doses. The tip mounting stage is made in such a way that it is electrically insulated from the cryostat, as an extremely high voltage is going to be applied via HV feedthrough, and also allows a high thermal conductivity for cooling the tip down to the liquid nitrogen temperature. This high thermal conductivity is achieved by attaching the tip mounting bracket to a piece of sapphire. A DVD or CCD camera is also needed to capture short movies and image frames from the FIM screen.

Normally, the tip used in this method can be electrochemically etched from a cheap polycrystalline tungsten wire in NaOH or KOH solutions; expensive single crystal wire can be used as well but is not necessary. The tip is then loaded in the UHV system and annealed to approximately 1000°C for cleaning oxide and other contaminants from it. Subsequently the tip is transferred to the FIM chamber and mounted on the cryostat in such a way that it is facing the screen at 3 cm apart. The tip is then cooled to the liquid nitrogen temperature and He gas is introduced at a pressure of 8×10^{-6} Torr. Very likely, the FIM image reveals a rough tip surface at the initial stages. The voltage should be increased carefully to field-evaporate the undesired protrusions and obtain an atomically smooth apex surface. For the tip presented in Fig. 7, the best imaging voltage for the clean surface was 5.9 kV. Then nitrogen gas is introduced at around 8×10^{-8} Torr, all gases being monitored by the mass spectrometer. Maintaining such conditions of gas levels and imaging fields lead to the etching and thus the removal of atoms strictly from the periphery of the tip apex and the tip shank. To avoid the evaporation of the atom at the center of the apex, the applied voltage is gradually decreased. The etching process continues as long as nitrogen gas is still in the chamber. With enough care this reaction can be controlled to form an ultrasharp tip. The tip sharpening can be recognized from the decrease of the tip apex and the increase of the resolution of the image. This is presented in the sequence of FIM images in Fig. 7, which were taken at different time intervals during the entire etching process. Tips with a few atoms (two atoms) and a single atom can be readily fabricated, as we see in Figs. 7(e) and 7(f). Once the single atom tip structure is attained, the nitrogen source has to be shut down and the tip voltage lowered slightly to avoid any further enhancement in the field at the tip end. The tip voltage at the last atom in Fig. 7(f) is 1.9 kV. For further illustration the tip-sharpening process and the decrease in the apex area are shown in the schematic in Fig. 8(a). The dotted line, dashed line and the gray area represent the initial, the intermediate and the final etched tip, respectively. The tip sharpening can be verified by sequentially removing the top atomic layers as in the schematic in Fig. 8(b). The initial nanotip, the intermediate field-evaporated tip and the final blunt tip are represented by the dotted line, the dashed line and the gray

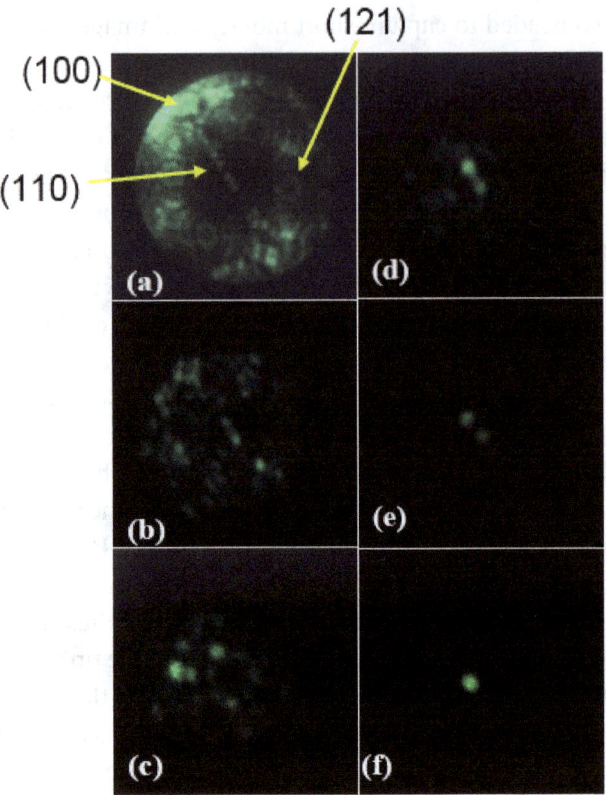

Fig. 7. Frames captured during the controlled nitrogen reaction with the tungsten FIM tip. (a) An atomically clean and broad tip obtained after field evaporation at the imaging voltage 5.9 kV. (b)–(f) Frames taken at different and successive stages of the nitrogen etching process. The apex area decreases until only two-atom and one-atom tip ends are obtained in (e) and (f), respectively. The imaging voltage was reduced gradually, from 5.9 kV at (a) to 1.9 kV at (f).

area. This tip destruction can be done by increasing the tip voltage back up gradually. The removal of the topmost atom of such single atom tips likely reveals an atomic layer of a few atoms, as we will see in the next section. The continuous destruction of the top atomic layers would give an estimate of the nanotip shape that has been formed. To ensure a relatively clean tip, the tip has to be annealed carefully prior to the sharpening process, so that no significant contamination would appear during the deconstruction process.

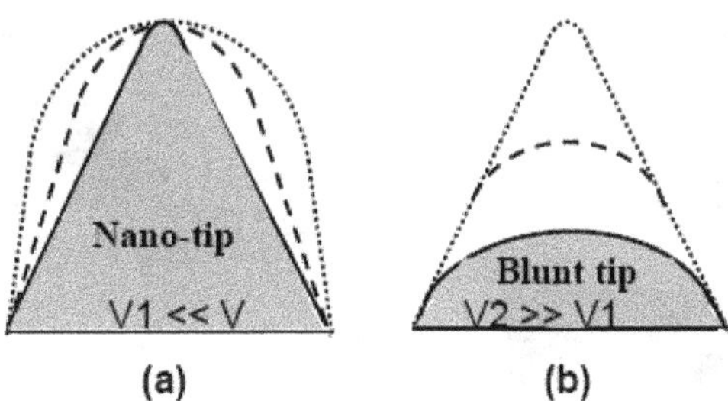

Fig. 8. (a) Sketch of successive tip sharpening stages during nitrogen etching. The dotted line refers to the original tip, the dashed line an intermediate etched tip, and the gray area the final stage (nanotip) at V1 $\ll$ V (in Fig. 6). (b) Sketch of the evaporation and flattening of the tip. The dotted line represents the initial nanotip, the dashed line an intermediate field-evaporated tip, and the gray area the final broadened tip at V2 $\gg$ V1.

1.7.2. *Tip apex modeling and nanotip reconstruction*

In the FIM, only the topmost apex atoms of a nanotip are projected on a 2D screen and it is rather difficult to get a side view of the tip. To estimate the overall tip shape underneath the last atom, the nanotip has to be deconstructed by removing the top atomic layers one by one. The removal of an entire top layer normally reveals the apex of an ultrasharp tip with a very small atomic structure. In fact, when only a few atomic layers (steps) can be imaged in the FIM with insufficient identified atomic planes, the conventional "ring counting" method for tip radius estimation, described in Sec. 4, cannot be used. Therefore, the most appropriate way of estimating the nanotip radius is to build a spherical crystal that can reproduce the location of surface atoms, and hence the surface atomic configuration on the apex. This can be done by using any of the commercially available crystal-building software programs. The initial step of this modeling process is to determine the orientation of the nanotip crystal and the azimuthal rotation, as the crystal structure is already known (i.e. bcc for tungsten). There are two ways of doing this — either from the original blunt tip, where the main planes can be identified, or after destroying the nanotip and removing enough atomic layers. Figures 9(a)–9(c) represent a deconstructed tip throughout the removal of the top five atomic layers. Figure 10(b) shows the tip after the removal of the topmost single atom, while Fig. 10(c) shows

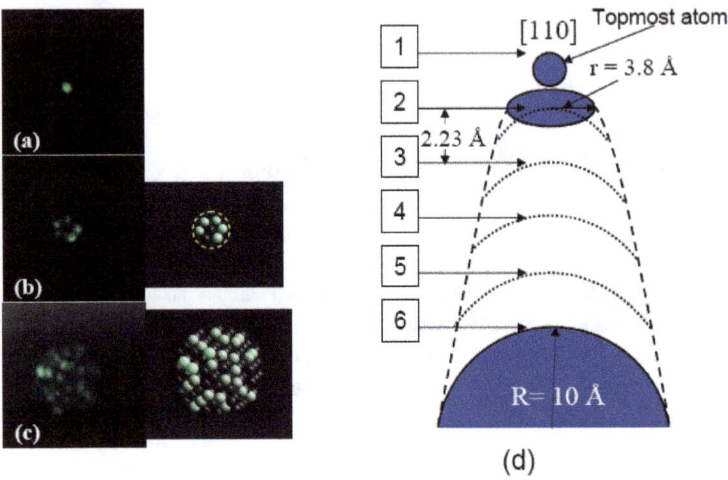

Fig. 9. FIM images captured during a subsequent field evaporation of a single atom tip in the absence of the reactant nitrogen gas. (a) A single atom tip at 1.9 kV just before starting the field evaporation process. (b) The successive atomic plane appeared after evaporation of the topmost atom. The next ball model is a (110) plane of five highlighted edge atoms and of radius 2.8 Å (c) A tip apex after successive removal of the top five atomic layers at 2.4 kV. The next ball model reproduces the atomic feature with a radius of 10 Å. (d) A schematic construction of the nanotip tip shape of six atomic layers along [110] and spaced by 2.23 Å.

the tip apex after the removal of the top five atomic layers at 2.4 kV. The orientation and azimuthal rotation of the crystal were determined from the broadest initial tip in Fig. 7(a), where the center (110) (largest) plane is clearly identified along with the other main planes like (121) and (100). The spherical crystal can be built by taking an appropriate spherical cut from relatively big cubic crystal. This cut is taken very carefully, so that the actual apex atomic arrangement, as it appears in the FIM image, can be reproduced in the corresponding model and most of the atoms can be located. Usually, the bright spots in the FIM image represent the edge atoms of a particular plane; the brighter ones are atoms protruding more from their lattice positions and are experiencing a stronger electric field and thus a surface reconstruction.[78,79] Therefore, these atoms are colored with different shades of green in the corresponding ball model. However, in the real tip not all of the atoms exist on their lattice positions on the apex or some of them do not have a strong electric field to be imaged in the FIM. Consequently,

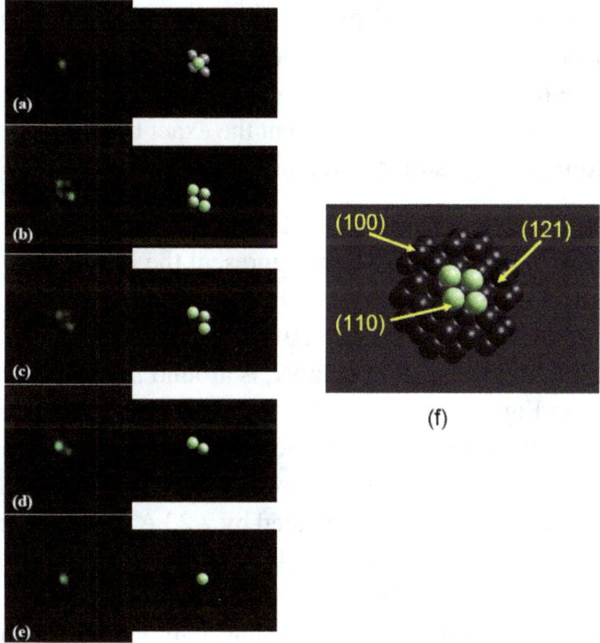

Fig. 10. (a) A single atom tip made from a polycrystalline W wire. (b) The removal of the single atom reveals an apex of four atoms. (b)–(e) A sequence of FIM images demonstrates, along with corresponding ball models, controlled removal of atoms one by one and the restoration of the single atom tip. (f) A spherical crystal of a (bcc) W lattice, with the same orientations of the original tip in Fig. 7(a), illustrating the extraction of the four-atom configuration in Fig. 10(b).

these atoms are either removed from the ball model or colored black. It is vital in this endeavor to find a unique match between the model and the image where increasing or decreasing the model size by 1 Å will result in severe disagreement between the model and the FIM image.

We have to consider in this modeling process two facts:

(i) The spacing between consecutive atomic layers (rings) in the spherical model as viewed from the top appears slightly smaller than that on the corresponding image. This is because the normal electric field on the surface is projecting the actual atomic spacing of all planes on the screen and resulting in full recognition, particularly for side planes. In contrast, the ball model represents the direct top view and therefore shows a smaller spacing, especially for side planes.

(ii) Surface atoms are often displaced from the regular bulk lattice positions and undergo surface reconstruction due to the presence of a strong electric field stress.[78,79] Unlike the FIM image, the positions of the surface atoms in the model represent the exact bcc lattice locations of these atoms, and no surface modification was made.

The ball models are presented next to FIM images in Fig. 9, except for the single atom tip as in Fig. 9(a), and represent the best match to the FIM atomic features. Figure 9(c) shows the tip apex after the removal of the five consecutive atomic layers from the (110) plane. The radius of the atomic tip apex in Fig. 9(a), the single atomic layer, is around 3.8 Å, while the radius of the nanotip in Fig. 9(c), as estimated from the next ball model, is 10 Å.

To estimate the overall shape of the initial nanotip (before its destruction), we visualize how the tip is reconstructed from all of the models stacked on top of each other and separated by 2.23 Å (the distance between two successive atomic layers along the [110] direction of the bcc W crystal); Fig. 9(d) illustrates the presumed nanotip shape.

There are several indications confirming a tip sharpening effect during the etching (sharpening) and deconstruction processes:

(i) The apex area decreases as the tip gets sharper and the dark area (low field region) around the apex increases (as in Fig. 7).
(ii) The increase in the image magnification goes as $M \propto 1/R$. This can be recognized clearly from the increase of the spot size and the spacing between spots in the consecutive FIM images during the sharpening process (Fig. 7).
(iii) The voltage drops from 5.9 kV for the initial broad tip to 1.9 kV for the single atom tip [Figs. 7(a)–7(f)].
(iv) The deconstruction of the tip by the field evaporation process reveals an extremely small tip apex underneath the topmost atom (Fig. 9).

1.7.3. *Controllability and reproducibility of the technique*

The described "controlled field assisted etching method" above can be applied to any tip crystal orientation as well as to polycrystalline wires, unlike other methods previously explored, which are limited to single crystals, mostly W(111). This special advantage is attributed to the fact that nitrogen adsorbs strongly on most of the low index W facets with different rates as long as the electric field on these sites is below the nitrogen ionization value. In Ref. 58 two cases of single atom tips fabricated from two different single crystal wires are demonstrated. The first single atom tip was made from (110) single crystalline tungsten with an initial tip radius of ~16 nm. The second single atom tip was fabricated from a single crystal W (310) wire with an initial radius of ~20 nm. The two single atom tips showed a similar FEM feature (a single round bright spot).

For various applications in surface science and nanotechnology utilizing scanning probe microscopy or electron microscopy, the tip robustness and reproducibility are important factors. For instance, when these tips are used in atomic lithography or for the manipulation of molecules and atoms, it is likely that the atomic structure of the apex will change, particularly the last atom end. Therefore, restoration of the single atom structure *in situ* with minimum effort is crucial for a reliable nanotip technology. In Fig. 10 we demonstrate step-by-step removal of the single atom tip and its restoration from the underneath four-atoms layer. Figure 10(a) shows the single atom just before it was field-evaporated. The sequence of images in Figs. 10(b)–10(e) demonstrates the sequential removal of these atoms one by one and stopping when the last atom is left and another single atom tip is formed in Fig. 10(e). The atomic configuration of a small atomic plane, like the one in Fig. 10(b), can be reproduced by building a ball model with the same top atomic arrangement. The initial and most important step in this endeavor is to determine the tip crystal orientation and azimuthal rotation. These decisive tip identities can be extracted by identifying the center plane and one of the side planes of the initial broad apex or after destroying the nanotip (via evaporating several top atomic layers); in both cases the tip has to be fixed in place throughout the entire experiment. Then a spherical model of a bcc structure and W lattice constant can be built and oriented to match the same actual blunt tip planes. The tip radius has to be selected in such a way that the top atomic structure is quite similar to the atomic arrangement with respect to the FIM image. The four-atom configuration in Fig. 10(b) can be reproduced by building a bcc spherical crystal with a lattice constant 2.23 Å and a radius of 7.7 Å and oriented to match the original tip planes, as depicted in Fig. 10(f).

We have to point out that the sequential removal of atoms takes place by the field enhancement due to nitrogen etching and not due to the conventional field evaporation, which requires an increase in the applied voltage. This sequential removal of atoms is illustrated in the models next to Figs. 10(b)–10(e).

1.8. Field Emission Characteristics of Single Atom Tips

Single atom tips show a distinct field emission feature (a round spot on the FEM screen),[46,50,58] which results from a highly collimated electron beam. This feature is a signature of the confined electron beam from the last single atom.[46,58] The coherence of such electron beams has been proven experimentally by placing a carbon nanotube a few nm away from the electron source and obtaining an interference pattern — a hologram — on the FEM screen.[46,56,57,80–87] Due to the low energy of the electron beam, single atom tips are ideal for low energy lensless projection microscopes.[56,80,81] The total energy distribution (TED) measurements from single atom tips show that the TED spectrum is composed of several separated peaks at different applied voltages[46] and none of these peaks are pinned at the bulk Fermi level, as is the case for macroscopic tips. This suggests the presence of a local band structure at the tip apex as the formation of a number of peaks occurs, because the emitted electron undergoes resonance tunneling through these bands.[46,83,87]

Another interesting feature of the single atom tip is the outstanding brightness value. "Brightness" is generally defined as the field emission current density divided by the solid angle of the electron beam. This parameter is a crucial factor in electron microscopy, particularly for the transmission electron microscope (TEM), because it determines the contrast and resolution of the image. In fact, ultrahigh resolution and atomic resolution can be achieved in a TEM if the electron virtual source size is small and the electron beam is highly collimated. This is achieved in existing TEMs by using a series of apertures and magnetic lenses. As we have mentioned earlier, the electron beam of a single atom tip is quite coherent and thus would help to avoid using such extra tools for collimating the electron beam. The brightness has been measured carefully by connecting a floating picoammeter in series between the high voltage power supply and the tip as the FEM pattern (i.e. a round bright spot) is being monitored on the screen. Assuming that all electrons are emitted from the last single atom and collected on the screen, a brightness value of around $B = 6.0 \times 10^9\,\mathrm{A\,cm^{-2}sr^{-1}}$ can be attained at a voltage of 250 V and an emission current of ~35 nA.[88]

This brightness value can be appreciated when compared with previous values for conventional electron sources as well as for nano and single atom sources. For conventional tips, with radii of ~100 nm, the typical brightness value is $\sim 10^5\,\mathrm{A\,cm^{-2}sr^{-1}}$.[43,44] Carbon nanotubes have brightness in the order of $1 \times 10^7\,\mathrm{A\,cm^{-2}sr^{-1}}$ when a few hundred volts are applied on the tip.[89,90] The brightness of single atom tips made by Binh and other researchers have values of $\sim 3 \times 10^8\,\mathrm{A\,cm^{-2}sr^{-1}}$ when similar voltage ranges are applied.[81,86] Therefore, the brightness of these single atom tips is at least one order of magnitude larger than the highest value so far. This entitles the single atom tip fabricated by the field-assisted etching method to be an ideal coherent and bright electron source for low energy lensless microscopes as well as for ultrahigh resolution transmission electron microscopes.

1.9. Applications of Nanotips in Scanning Probe Microscopy and Future Trends

In scanning probe microscopes the tip end apex shape and size are direct factors of the resolution. For instance, in STMs, the smaller the tip apex the smaller the tunneling area and hence the higher the spatial resolution. In the case of rough tips with multiple protrusions, repeated features and distorted structures may be seen in the image. This tip roughness can be readily seen in the initial stage of FIM imaging and during the field increase and hence evaporation process. Very likely, the ultimate resolution, i.e. atomic resolution, is attained in STMs when one protrusion is higher than the neighboring ones by a few angstroms. This is due to the fact that the tunneling current decays exponentially with the distance between the tip and the substrate surface. Therefore, the image resolution is determined by the highest protrusion. For the STM, the advantage of fabricating nanotips is that an immediate atomic resolution is obtained as the tip is made of only one sharp end with an atomic apex.[52] Moreover, nanotips in STM machines are needed to manipulate atoms and molecules in a reliable way. Early successes were obtained in 1990s by D. Eigler *et al.*,[30] who were able to construct a patterned array of xenon atoms on a nickel (110) surface, and by Avouris *et al.*,[91] who were able to manipulate atoms and nanoclusters on a Si(111) surface. Nevertheless, the development of this work is still slow, due to the lack of reliable technologies for producing extremely sharp tips and restoring them *in situ* upon losing the last atom at the end of the tip. Researchers are also interested in forming multiple electronic contacts with a single molecule or a nanostructure.[92–95] These endeavors are confronted by the fact that the tip apex size and shape are very likely to be bigger than the object to be contacted. The tip shape also limits the tip to tip spacing and hence the number of tips that can converge on a single molecule.[96] Therefore, ultrasharp tips with a high aspect ratio are highly required for making reliable single and multiple contacts with a nano-object, as well as for atomic and molecular manipulation.[97] In atomic force microcopy the resolution is determined by the long range interaction with the macroscopic tip. Thus, well-defined and high-aspect-ratio sharp tips are required for obtaining optimum resolution in the AFM. In light of these facts we foresee that the field-assisted etching method will provide the best solution for improving the performance of scanning probe microscopes and escalate the development of atomic and molecular technologies. This augments their interesting applications as super electron and ion sources.[45,56–58,80–82,98–105]

For reliable nanotip technology it is essential that the very sharp tip end can be recovered after air exposure. This is indeed required when those tips have to be shipped to other machines where they can be used, like the STM, AFM, TEM or SEM. Fortunately, it was proven that single atom tip structure could be restored after a few days of air exposure by a gentle annealing procedure, with the overall nanotip still intact.[52a]

1.10. Conclusion

The various common and previously known tip-sharpening methods have been briefly reviewed and some limitations that are impeding the development of such technologies have been summarized. For instance, in the deposition method the atoms are projected onto the tip surface with insufficient control on the number of atoms or on the adsorption sites. The field–surface melting and "build-up" methods very likely involve other adsorbates on the tip shank which migrate to the apex during the field increase, in addition to the fact that the formation of a single atom tip cannot be determined directly during the process unless the tip is characterized again in an FIM. Similar concerns are applicable to the Pd-covered W(111) tip. To appreciate the recent (field-assisted nitrogen etching) technique, it is worth highlighting some of the advantages that are inherent in the method:

(i) The single atom tip can be readily and reproducibly formed *in situ* and the whole process is monitored; atom-by-atom modification is possible.
(ii) This process avoids the complexities of the heat treatment and the deposition of external atoms, as the single atom tip is formed from the natural atomic metal structure.
(iii) More importantly, this recent method can be applied to tips of any crystal orientation as well as to polycrystalline wires. Therefore, it would provide a good solution to the problems that restrict the applications of scanning probe microscopes and electron microscopes in nanotechnology.

Acknowledgments

The authors would like to thank Ma Han Thu Lwin for her valuable assistance in setting up the field emission and ion microscopes. This work was supported by the Agency for Science, Technology and Research (A*STAR) of Singapore through the VIP "Atomic Scale Technology" program.

References

1. G. Fursey, *Field Emission in Vacuum Microelectronics* (Kluwer/Plenum, New York, 2005).
2. E. W. Muller and T. T. Tsong, *Field Ion Microscopy: Principles and Applications* (American Elsevier, 1969).
3. A. J. Melmed, *Appl. Surf. Sci.* **94/95**, 17 (1996).

4. A. J. Melmed and E. W. Müller, *J. Chem. Phys.* **29**, 1037 (1958).
5. W. R. Graham, F. Hutchinson and D. A. Reed, *J. Appl. Phys.* **44**, 5155 (1973).
6. P. D. Cobden, B. E. Nieuwenhuys and V. V. Gorodetskii, *Appl. Catal. A* **188**, 69 (1999).
7. E. W. Muller and K. D. Rendulic, *Science* **156**, 961 (1967).
8. T. Al-kassab, M. P. Match and H. Wollenberger, *Appl. Surf. Sci.* **87/88**, 329 (1995).
9. T. Y. Fu, L. C. Cheng, Y. J. Hwang and T. T. Tseng, *Surf. Sci.* **507–510**, 103 (2002).
10. H. Morikawa, M. Kokura, F. Iwatsu and T. Terao, *Thin Solid Films* **254**, 103 (1995).
11. A. Szczepkowicz and A. Ciszewski, *Surf. Sci.* **515**, 441 (2002).
12. A. Szczepkowicz and A. Ciszewski, *Vacuum* **54**, 89 (1999).
13. L. E. Muller, O. T. Inal and H. P. Singh, *Thin Solid Films* **9**, 241 (1972).
14. R. B. Sharma, P. W. Yawalkar, N. Pradeep and D. S. Joag, *Ultramicroscopy* **73**, 99 (1998).
15. D. W. Bassett, *Thin Solid Films* **48**, 237 (1978).
16. E. W. Muller, *Phys. Rev.* **102**, 618 (1956).
17. A. L. Suvorov, Y. N. Cheblukov, A. F. Bobkov, S. V. Zaitsev, S. V. Latishev, E. N. Skorokhodov and A. E. Stepanov, *Mater. Sci. Eng.* **A270**, 94 (1999).
18. Z. Z. Xiang, Z. G. Min, D. Min, J. X. Xi, H. S. Min, S. J. Ping, G. Z. Nan, Z. X. Yu, L. W. Min, W. J. Lei and X. Z. Quan, *Chin. Phys.* **11**, 804 (2002).
19. Z. Knor, S. Biehl, J. Plsek, L. Dvorak and C. Edelmann, *Vacuum* **51**, 11 (1998).
20. M. Wohlmuth and E. Bechtold, *Appl. Surf. Sci.* **5**, 243 (1998).
21. T. T. Tsong, *Chin. J. Phys.* **35**, 888 (1997).
22. N. Ernst, *Surf. Sci.* **200**, 275 (1988).
23. G. L. Kellogg, *Surf. Sci. Rep.* **21**, 1 (1994).
24. M. K. Miller, A. Cerezo and G. D. W. Smith, *Atom Probe Field Ion Microscopy* (Clarendon, Oxford, 1996).
25. T. Tsong, *Atom-Probe Field Ion Microscopy* (Cambridge University Press, New York, 1990).
26. T. T. Tsong, *Surf. Sci.* **299/300**, 153 (1993).
27. T. T. Tsong, H. M. Liu, Q. J. Gao, Y. Liou and D. L. Feng, *Surf. Sci.* **200**, 220 (1988).
28. J. J. Langer and M. Martyński, *Synthetic Metals* **107**, 1 (1999).
29. G. P. Lopinski, D. D. M. Wayner and R. A. Wolkow, *Nature* **406**, 48 (2000).
30. D. M. Eigler and E. K. Schweizer, *Nature* **344**, 524 (1990).
31. M. Kanai, T. Kawai, K. Motai, X. D. Wang, T. Hashizume and T. Sakura, *Surf. Sci. Lett.* **329**, L619 (1995).
32. P. G. Piva, G. Dilabio, J. L. Pitters, J. Zikovsky, M. Rezeq, S. Dogel, W. A. Hofer and R. A. Wolkow, *Nature* **435**, 658 (2005).

33. L. Grill, K. H. Rieder, F. Moresco, G. Rapenne, S. Stojkovic, X. Bouju and C. Joachim, *Nat. Nanotechnol.* **2**, 95 (2997).
34. J. Yang, J. Deng, N. Chandrasekhar and C. Joachim, *J. Vac. Sci. Technol. B* **25**, 1694 (2007).
35. A. Pomyalov and Y. Manassen, *Surf. Sci.* **382**, 275 (1997).
36. J. B. Park, B. Jaeckel and B. A. Pakinson, *Langmuir* **22**, 5334 (2006).
37. S. Hasegawa, I. Shiraki, F. Tanabe and R. Hobara, *Curr. Appl. Phys.* **2**, 465 (2002).
38. F. Rochet, G. Dufour, H. Roulet, N. Motta, A. Sgarlata, M. N. Piancastelli and M. D. Crescenzi, *Surf. Sci.* **319**, 10 (1994).
39. W. Haiss, R. J. Nichols, H. V. Zaling, S. J. Higgins, D. Bethell and D. J. Schiffrin, *Phys. Chem. Chem. Phys.* **6**, 4330 (2004).
40. P. Avouris, *Acc. Chem. Res.* **28**, 95 (1995).
41. W. Haiss, H. V. Zalinge, H. Höbenreich, D. Benthell, D. J. Schiffirn, S. J. Higgins and R. J. Nichols, *Langmuir* **20**, 7694 (2004).
42. C. Joachim, J. K. Gimzewski, R. R. Schlittler and C. Chavy, *Phys. Rev. Lett.* **74**, 2102 (1995).
43. P. Doumergue, L. Pizzagalli, C. Joachim, A. Altibelli and A. Baratoff, *Phys. Rev. B* **59**, 15910 (1999).
44. D. Natelson, L. H. Yu, J. W. Ciszek, Z. K. Keane and J. M. Tour, *Chem. Phys.* **324**, 267 (2006).
45. H. J. Gao, K. Sohlberg, Z. Q. Xue, H. Y. Chen, S. M. Hou, L. P. Ma, X. W. Fang, S. J. Pang and S. J. Pennycook, *Phys. Rev. Lett.* **84**, 1780 (200).
46. V. T. Binh, S. T. Purcell, V. Semet and F. Feschet, *Appl. Surf. Sci.* **130–132**, 803 (1998).
47. V. T. Binh, *Surf. Sci. Lett.* **202**, L539 (1988).
48. H. W. Fink, *IBM J. Res. Dev.* **30**, 461 (1986).
49. T. Y. Fu, L. C. Cheng, C. H. Nien and T. T. Tsong, *Phys. Rev. B* **64**, 113401 (2001).
50. H. S. Kuo, I. S. Hwang, T. Y. Fu, J. Y. Wu, C. C. Cheng and T. T. Tsong, *Nano Lett.* **4**, 2379 (2004).
51. K. Nagaoka, H. Fujiii, K. Matsuda, M. Komaki, Y. Murata, C. Oshima and T. Sakurai, *Appl. Surf. Sci.* **182**, 12 (2001).
52. (a) M. Rezeq, J. Pitters and R. Wolkow, *J. Chem. Phys.* **124**, 204716 (2006); (b) M. Rezeq, J. Pitters and R. Wolkow, *United States Patent No.* 7,431,856 (2008).
53. J. Unger, Y. A. Vlasov and N. Ernst, *Appl. Phys. Lett.* **87/88**, 45 (1995).
54. A. Knoblauch, C. Wilbertz, T. Miller and S. Kalbitzer, *J. Phys. D* **29**, 470 (1996).
55. M. J. Francen, M. H. F. Overwijk and P. Kruit, *Appl. Surf. Sci.* **146**, 357 (1999).
56. H. W. Fink, W. Stocker and H. Schmid, *Phys. Rev. Lett.* **65**, 1204 (1990).
57. H. Schmid and H. W. Fink, *Nanotechnology* **5**, 26 (1994).
58. M. Rezeq, J. Pitters and R. Wolkow, *J. Scan. Probe Microsc.* **2**, 1 (2007).

59. P. H. Culter, J. He, J. Miller, N. M. Miskovsky, B. Weiss and T. E. Sullivan, *Prog. Surf. Sci.* **42**, 169 (1993).
60. K. Yuasa, A. Shimoi, I. Ohba and C. Oshima, *Surf. Sci.* **520**, 18 (2002).
61. R. G. Fobers, *Ultramicroscopy* **79**, 11 (1999).
62. C. J. Edgcombe, *Ultramicroscopy* **95**, 49 (2003).
63. I. Brodie and C. A. Spindt, in *Advances in Electronics and Electron Physics*, ed. P. W. Hawkes (Academic, San Diego, 1992), Vol. 83, p. 11.
64. V. T. Binh and Garcia, *Ultramicroscopy* **44–44**, 80 (1992).
65. J. T. Yates, Jr., R. Klein and T. E. Madey, *Surf. Sci.* **58**, 469 (1976).
66. T. E. Madey, J. T. Yates, Jr. and N. E. Erickson, *Surf. Sci.* **43**, 526 (1974).
67. K. Ota and S. Usami, *Surf. Sci.* **287/288**, 99 (1993).
68. M. Serrano and G. R. Darling, *Surf. Sci.* **532–535**, 206 (2003).
69. T. A. Delchar and C. Ehrlich, *J. Chem. Phys.* **42**, 2686 (1964).
70. P. W. Tamm and L. D. Schmidt, *Surf. Sci.* **26**, 286 (1971).
71. L. Holmgren, M. Anderson and A. Rosén, *J. Chem. Phys.* **109**, 3232 (1998).
72. J. T. Yates, Jr. and T. E. Madey, *J. Chem. Phys.* **43**, 1055 (1965).
73. T. E. Madey and T. Yates, Jr., *J. Chem. Phys.* **44**, 1675 (1965).
74. K. Rendulic and Z. Knor, *Surf. Sci.* **7**, 205 (1967).
75. J. R. Hiskes, *Phys. Rev.* **122**, 1207 (1960).
76. G. E. Ehrlich and F. G. Hudda, *J. Chem. Phys.* **36**, 3233 (1962).
77. T. Schimizu, A. Ohi and H. Tokumoto, *Surf. Sci.* **429**, 143 (1999).
78. M. K. Miller, *Atom Probe Tomography: Analysis at the Atomic Level* (Kluwer/Plenum, New York, 2000), p. 52.
79. R. G. Forbes, *J. Phys. D* **18**, 973 (1985).
80. H. W. Fink and H. Schmid, *Phys. Rev. Lett.* **67**, 1543 (1991).
81. V. T. Binh and V. Semet, *Ultramicroscopy* **73**, 107 (1998).
82. H. W. Fink, W. Stocker and H. Schmid, *J. Vac. Sci. Technol. B* **8**, 1323 (1990).
83. V. T. Binh, S. T. Purcell, N. Garcia and J. Doglioni, *Phys. Rev. Lett.* **69**, 2527 (1992).
84. V. T. Binh and S. T. Purcell, *Appl. Surf. Sci.* **111**, 157 (1997).
85. J. C. H. Spence, W. Qian and M. P. Silverman, *J. Vac. Sci. Technol. A* **12**, 542 (1994).
86. M. P. Siverman, W. Strange and J. C. H. Spence, *Am. J. Phys.* **63**, 800 (1995).
87. V. T. Binh, S. T. Purcell, G. Gardet and N. Garcia, *Surf. Sci. Lett.* **279**, L197 (1992).
88. M. Rezeq, C. Christian and N. Chandrasekhar, *Surf. Sci.* **603**, 697 (2009).
89. N. D. Jonge, Y. Lamy, K. Schoots and T. Oosterkamp, *Nature* **420**, 393 (2002).
90. K. Teo, *JOM* **59**, 29 (2007).
91. P. Avouris, I. W. Lyo and Y. Hasegawa, *J. Vac. Sci. Technol. A* **11**, 11725 (1993).
92. C. Joachim, J. K. Gimzewski and A. Aviram, *Nature* **408**, 541 (2000).
93. C. A. Stafford, D. M. Cardamone and S. Mazumdar, *Nanotechnology* **18**, 424014 (2007).
94. C. Joachim, J. K. Gimzewski, R. R. Schlittler and C. Chavy, *Phys. Rev. Lett.* **74**, 2102 (1995).

95. J. Reichert, and D. Beckmann, H. B. Weber, M. Mayor and H. V. Löhneysen, *Phys. Rev. Lett.* **29**, 176804 (2002).
96. F. Moresco, L. Gross, M. Alemani, K.-H. Rieder, H. Tang, A. Gourdon and C. Joachim, *Phys. Rev. Lett.* **91**, 36601 (2003).
97. J. A. Stroscio and R. J. Celotta, *Science* **306**, 242 (2004).
98. J. C. H. Spence, *Mater. Sci. Eng.* **R26**, 1 (1999).
99. D. C. Joy, Y. S. Zhang, X. Zhang, T. Hashimoto, R. D. Bunn, L. Allard and T. A. Nolan, *Ultramicroscopy* **51**, 1 (1993).
100. S. T. Purcell and V. T. Binh, *Appl. Phys. Lett.* **75**, 1332 (1999).
101. W. Qian, M. R. Scheinfein and J. C. H. Spence, *J. Appl. Phys.* **73**, 7041 (1993).
102. P. Hommelhoff, Y. Sortais, A. A. Talesh and M. A. Kasevich, *Phys. Rev. Lett.* **96**, 77401 (2006).
103. E. Rokuta. T. Itagaki, T. Ishikawa, B. L. Cho, H. S. Kuo, T. T. Tsong and C. Oshima, *Appl. Surf. Sci.* **252**, 3686 (2006).
104. J. C. H. Spence, *Micron* **28**, 101 (1997).
105. L. Livadaru, J. Mutus and R. A. Wolkow, *Ultramicroscopy* **108**, 472 (2008).

Moh'd Rezeq obtained a PhD in Condensed Matter Physics from the University of Ottawa, Canada, in 2002. He then joined the National Institute for Nanotechnology (NINT) in Edmonton, Alberta, Canada, as a researcher. He used the scanning tunneling microscope (STM) to investigate the electric characteristics of single molecules and nanometal–semiconductor contacts. Employing the field ion microscope (FIM), he developed a new technology for fabricating ultrasharp tips with a radius of <1 nm. These nanotips have crucial applications in atomic scale microscopy and lithography. From 2006 to 2009, Dr Rezeq was a Senior Research Engineer at the Institute of Materials Research and Engineering (IMRE), Singapore. There he conducted research in nanotechnology and constructed the first FIM at IMRE. He was also in charge of the molecular-interconnect project. In August 2009, Dr Rezeq joined Khalifa University of Science, Technology and Research (KUSTAR), UAE, as an assistant professor in Physics. He is continuing his research in nanotechnology, with the focus on its applications in microscopy, medicine, electronics and bioengineering.

Christian Joachim is First Class Director of Research at CNRS, head of the nanoscience group at CEMES/CNRS (www.cemes.fr/GNS/) and Adjunct Professor of Quantum Physics at Sup'Aero (ISAE Toulouse). He is A*STAR VIP at IMRE to develop atomic scale technology in Singapore and head of the WPI MANA-NIMS satellite in Toulouse. He coordinated the EU projects "Bottom-up Nanomachines" and "Pico-Inside." He is currently in charge of the Midi-Pyrenees research effort in Nanoscience (CPER G. Dupouy). He has directed two NATO Advanced Research Workshops on Nanoscale Sciences, in the early 1990s. Dr Joachim is the author of 200 scientific publications and has presented over 250 invited talks on electron transfer through a molecule, STM and atomic force microscopy (AFM) image calculations, tunnel transport through a molecule, molecular devices, nanolithography and single molecule machines. His book *Nanosciences: The Invisible Revolution* (World Scientific, 2009) presents the history of nanoscience and its political drawbacks to the general public.

In Situ STM Studies of Molecular Self-Assembly on Surfaces

Chapter **2**

Wei Chen*,† and Andrew T. S. Wee*

Abstract

Molecular self-assembly on surfaces via weak but selective noncovalent interactions offers a promising bottom-up approach to fabricating molecular nanostructure arrays with desired functionalities over macroscopic areas. In this chapter, we highlight recent progress in the fabrication of self-assembled molecular nanostructures on surfaces and surface nanotemplates, as investigated by *in situ* scanning tunneling microscopy. We describe the formation of various C_{60} nanostructure arrays on molecular surface nanotemplates of monolayer and bilayer α-sexithiophene nanostripes on Ag(111) and graphite, p-sexiphenyl nanostripes and pentacene networks on Ag(111), as well as the formation of two-dimensional supramolecular chiral networks of a binary molecular system involving pentacene and 3,4,9,10-perylene-tetracarboxylic-dianhydride on Ag(111). These novel organic nanojunction arrays have promising applications for future molecular-electronic or organic solar cell devices.

2.1. Introduction

The control and manipulation of supramolecular arrangements of molecules into highly periodic nanostructure arrays or well-defined organic thin films

*Department of Physics, National University of Singapore, 2 Science Drive 3, Singapore, 117542.
Emails: phycw@nus.edu.sg (W. Chen) and phyweets@nus.edu.sg (A. T. S. Wee)
†Department of Chemistry, National University of Singapore, 3 Science Drive 3, Singapore 117543.

on surfaces is a major scientific challenge, with potential applications in the fields of molecular and organic electronics, organic solar cells, solid state quantum computation and biosensors.[1] Examples of such applications are: (a) precisely addressing individual molecular devices for the realization of molecular electronics, which requires accurate prearrangement of functional molecules into well-defined device arrays;[2–4] (b) assembly of endohedral fullerenes with incarcerated electron spins into two-dimensional arrays on surfaces for fullerene-based quantum computers;[5] (c) patterning and immobilizing proteins or DNA at the nanometer scale to form well-ordered biomolecular superstructures for biosensors and controlled cell adhesion;[6–9] (d) growth of organic thin films with well-defined supramolecular arrangements for organic thin film transistors (OTFTs).[10]

Molecular self-assembly on surfaces or surface nanotemplates via weak but selective noncovalent interactions (including electrostatic forces, hydrogen bonds, van der Waals forces and metal–ligand interactions) offers a promising bottom-up approach to fabricating molecular nanostructure arrays with desired functionalities over macroscopic areas.[1,11–14] Such bottom-up approaches via self-assembly can be categorized into two groups: (i) self-assembly of functional molecules on surface nanotemplates with prepatterned and well-ordered preferential adsorption sites to selectively accommodate guest adsorbates (molecules or atoms); (ii) surface-supported one- or two-dimensional highly periodic supramolecular assemblies formed by directional and selective noncovalent intermolecular interactions, such as hydrogen bonding and metal–ligand interactions.

2.1.1. *Self-assembly on surface nanotemplates or nanostructured surfaces*

The controlled positioning and assembly of functional molecules on surfaces crucially depends on the interplay between various chemical and physical bonds formed and lateral interactions of different strengths and length scales.[1,11–16] Surface nanotemplates with periodic preferential adsorption sites can be used to selectively accommodate molecules to form molecular nanostructure arrays with periodicity matching the underlying nanotemplates.[1,11,12] There are many recently published works on the investigation of molecular self-assembly on various surface nanotemplates that are naturally or artificially patterned at the nanoscale. Examples are the use of surface reconstructions,[17–26] step bunching on vicinal surfaces,[27–30] adsorbate-induced surface superstructures,[31] strain relief patterns generated by the deposition of material with a different lattice constant to that of the substrate,[32–35] organized supramolecular architectures stabilized by directional noncovalent intermolecular interactions predominantly involving hydrogen bonds,[36–54] and metal–ligand interactions.[55–66] Among them, 2D surface nanoporous or nanocavity networks have attracted much attention. Examples are the formation of regular C_{60} arrays on a 2D nanoporous honeycomb supramolecular network on Ag-passivated Si(111) substrates stabilized through triple hydrogen bond coupling between perylene tetracarboxylic-di-imide (PTCDI) and melamine molecules;[36] C_{60} and biomolecules such as cystine and LL-diphenylalanine reversibly decorating highly ordered nanocavities formed by the self-assembly of Fe atoms and trimesic acid molecules via metal–ligand interactions on Cu(100);[57] and hexagonal C_{60} molecular dot array formation on a 2D porphyrin-based porous network[67] or a C_{60} nanomesh.[68]

2.1.2. *Self-assembled 2D molecular nanostructures via directional noncovalent or covalent intermolecular interactions*

Directional and selective intermolecular interactions, such as hydrogen bonding, metal–ligand interactions, as well as covalent bonding, have been widely used in supramolecular chemistry to fabricate engineered molecular crystals and, more recently, surface-supported 1D or 2D highly periodic supramolecular assemblies with good structural stability, including molecular supergratings,[42–44] honeycomb networks[36,48,57,62–68] and various rationally designed molecular nanostructures.[49–54] An elegant example is the construction of metal–ligand coordination 2D supramolecular nanocavities on metal surfaces with precisely engineered shape, symmetry and size by varying the ligand molecules as well as the relative concentration of the metal atoms and ligand molecules.[62–66] The hydrogen-bonded single-component, binary or multicomponent supramolecular assemblies on surfaces represent another versatile strategy for the fabrication of molecular nanostructures with a high degree of controllability and tunability. Examples are the highly directional 1D chiral chains of 4-[trans-2-(pyrid-4-vinyl)]benzoic acid (PVBA) on Ag(111)[43] and short molecular rods of adenine on Cu(110),[69] molecular nanoclusters of 1-nitronaphthalene on reconstructed Au(111),[70,71] extended 2D chiral networks of tartaric acid on Cu(111),[72] a large homomolecular porous network at a Cu(111)[48] and so on.

In this chapter, we review recent work in the fabrication of tunable C_{60} nanostructure arrays on various molecular surface nanotemplates and hydrogen-bonded 2D supramolecular chiral networks of binary molecular systems, studied using an *in situ* scanning tunneling microscope (STM).

2.2. *In Situ* Ultrahigh Vacuum Scanning Tunneling Microscopy

The molecular self-assembly processes on surfaces can be monitored in real time and space using an *in situ* STM. Figure 1 shows a typical multichamber ultrahigh vacuum (UHV) system employed for such *in situ* experiments.[73–75] The LT-STM can be cooled down to 5 K (liquid He) or 77 K (liquid N_2) for different experimental purposes. In order to achieve atomic resolution, all STM imaging was performed at 77 K to suppress molecular diffusion during scanning. The LT-STM is connected to an organic molecular beam deposition (OMBD) chamber with several low-temperature Knudsen cells for *in situ* deposition of organic molecules, a quartz crystal microbalance (QCM) for real-time deposition rate monitoring, and an ion sputter gun for sample cleaning.

The purity of the organic sources plays a crucial role in the fabrication of self-assembled molecular nanostructures. Prior to deposition, all molecules were further purified once or twice by gradient vacuum sublimation.

2.3. Self-Assembled C_{60} Nanostructures on Molecular Surface Nanotemplates

In this section, we describe three examples of molecular self-assembly on surface nanotemplates. The first example demonstrates the preferential

Fig. 1. The LT-STM setup at the NUS Surface Science Laboratory. The left chamber houses an Omicron LT-STM operated at L-He or L-N_2 temperature; the right chamber is a custom-designed OMBD system.

trapping of individual C_{60} molecules into the nanocavities of a C_{60} nanomesh, resulting in the formation of ordered 2D C_{60} arrays with a large intermolecular distance (2.10 nm) between the nearest neighbor C_{60} molecules.[68] As shown in Fig. 2(a), deposition of 0.7 ML pentacene on Ag(111) at room temperature (RT) followed by annealing at 380 K leads to the formation of a highly ordered pentacene molecular array with a rectangular unit cell of $\mathbf{a} = 2.90 \pm 0.02$ nm and $\mathbf{b} = 1.03 \pm 0.02$ nm. Each rodlike feature represents a single pentacene molecule. Depositing 0.7 ML C_{60} onto this pentacene superstructure at RT and subsequently annealing to 360 K gives rise to the formation of an extended 2D network with well-ordered nanocavities with a unit cell of $\mathbf{a}_1 = 2.62 \pm 0.02$ nm, $\mathbf{b}_1 = 2.10 \pm 0.02$ nm and $\boldsymbol{\alpha}_1 = 60°$ [Fig. 2(b)]; this is referred to as the

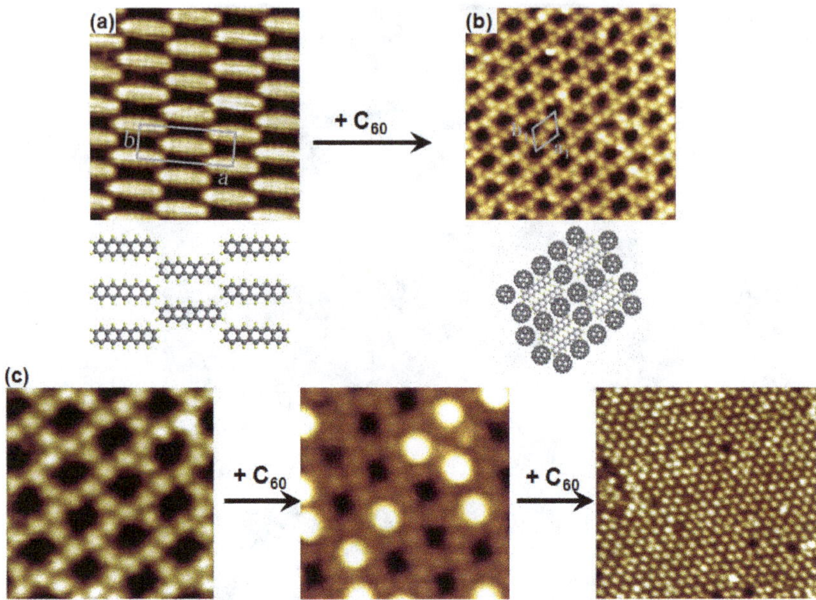

Fig. 2. (a) Top: STM image ($7 \times 7\,\text{nm}^2$; $V_{tip} = -0.9\,\text{V}$) of highly periodic pentacene superstructure on Ag(111). Bottom: Schematic model for the molecular packing structure of the pentacene structure. (b) Top: Typical STM image ($20 \times 20\,\text{nm}^2$; $V_{tip} = -1.5\,\text{V}$) of a C_{60} nanomesh. Bottom: Corresponding proposed model of this C_{60} nanomesh. (c) Sequential STM images showing the preferential trapping of individual C_{60} molecules into the nanocavities of the C_{60} nanomesh, leading to the formation of regular C_{60} arrays. (*Adapted from Ref. 68.*)

C_{60} nanomesh. This C_{60} nanomesh serves as an effective surface nanotemplate for trapping single C_{60} molecules in its nanocavities, as shown in Fig. 2(c). Upon further increasing the C_{60} coverage to 0.4 ML [Fig. 2(c)], C_{60} molecules almost completely occupy the nanocavities of the nanomesh template, forming an ordered 2D C_{60} molecular array with an intermolecular distance of about 2.10 nm. We propose that the topographic features of the C_{60} nanomesh create preferential sites for the immobilization of C_{60} molecules within the nanocavities, thereby forming an ordered 2D molecular array.

The second example is the formation of various C_{60} nanostructure arrays on molecular surface nanotemplates of monolayer and bilayer α-sexithiophene (6T) nanostripes on Ag(111) and graphite.[76–78] As shown in Figs. 3(a) and 3(b), 6T molecules can assemble into highly ordered

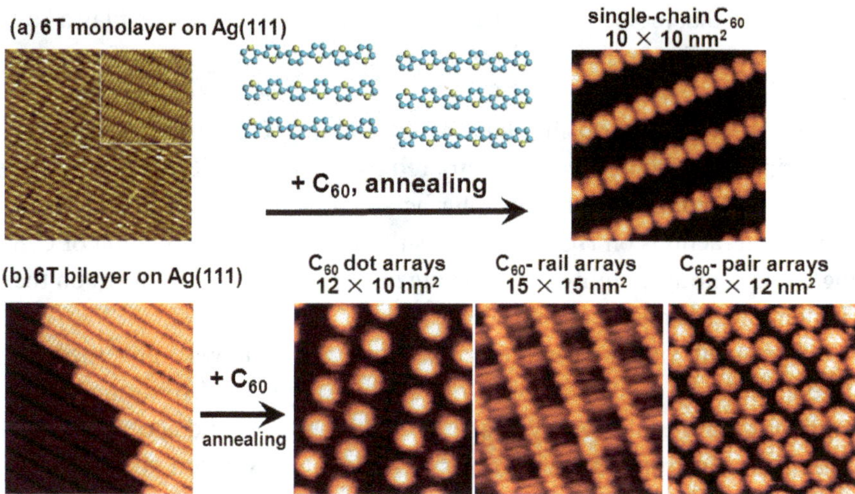

Fig. 3. (a) C_{60} single-chain arrays form on 6T monolayer nanostripes on Ag(111). (b) Tunable C_{60} nanostructure arrays including C_{60} dot arrays, rail-like arrays and pair arrays form on the 6T bilayer nanostripes on Ag(111). (*Adapted from Refs. 76 and 77.*)

monolayer or bilayer nanostripe arrays on Ag(111), via side-by-side packing, with their thiophene rings oriented parallel to the substrate surface. Both monolayer and bilayer 6T nanostripes serve as nanotemplates that direct the assembly of C_{60} molecules into various molecular nanostructure arrays. It is worth noting that the 6T monolayer nanostripes interact strongly with the underlying Ag substrate. The strong coupling between the 6T π-electrons and the Ag d-band ensures the structural rigidity of the 6T monolayer nanostripes during C_{60} deposition and postannealing, leading to the formation of long-range-ordered C_{60} single-chain arrays with an interchain distance of 2.31 ± 0.02 nm, as shown in Fig. 3(a).[76] In contrast, the coupling between the top 6T layer of the bilayer nanostripes and Ag(111) is attenuated by the underlying 6T monolayer. Therefore, the 6T bilayer nanostripes are mainly sustained by weak intermolecular van der Waals forces. This can easily lead to structural rearrangement of the underlying 6T layers during the deposition of C_{60} by careful control of the C_{60} coverage as well as postannealing temperature, thereby giving rise to the formation of tunable C_{60} nanostructure arrays including C_{60} dot arrays, rail-like arrays and pair arrays, as shown in Fig. 3(b).[77] These experiments demonstrate that molecule–substrate interfacial interactions also play a key role in assembling molecules into desired nanostructures on surface nanotemplates.

We further confirm this idea by assembling C_{60} on 6T monolayer nanostripes on graphite or highly ordered pyrolytic graphite (HOPG).[78] In this case, the molecule–substrate interfacial interactions are now dominated by π–π interactions on HOPG. As a result, instead of the formation of C_{60} single-chain arrays, highly ordered "zigzag" C_{60} chain arrays form on the 6T monolayer nanostripes on HOPG.[78]

It is well known that molecular assembly is largely governed by the combination of interactions with the underlying substrate and with neighboring molecules. As typical electron donor (6T) and acceptor (C_{60}) molecules, electron transfer from 6T to C_{60} is expected to dominate the C_{60}-6T intermolecular interactions, i.e. donor–acceptor intermolecular interactions. Hence, the formation of the long-range-ordered C_{60} nanostructure arrays on 6T nanostripes on Ag(111) or HOPG is proposed to arise from the delicate balance between the homo-intermolecular (C_{60}–C_{60} and 6T–6T, van der Waals forces), hetero-intermolecular (C_{60}–6T, charge transfer or donor–acceptor interaction) and molecule–substrate (π–d interaction for 6T–Ag and π–π interaction for 6T–HOPG) interfacial interactions.

The third example is the construction of well-ordered organic donor/acceptor nanojunction arrays comprising *p*-sexiphenyl (6P) and C_{60} via self-assembly of C_{60} on 6P nanostripes on Ag(111).[79] Analogous to the packing structure of the 6T monolayer on Ag(111), Fig. 4(a) shows the formation of well-ordered unidirectional 6P nanostripes on Ag(111) after the deposition of 1 ML 6P at RT and subsequent annealing at 380 K. Each rodlike bright feature in the STM image represents a single 6P molecule. The width of the 6P monolayer nanostripes is 2.95 ± 0.02 nm, and the periodicity along the nanostripe is 0.70 ± 0.02 nm. The formation of 6P monolayer nanostripes on Ag(111) can be understood in terms of self-assembly via side-by-side packing with extended molecular π-planes parallel to the Ag(111) surface, as shown by the schematic in Fig. 4(a), and this is referred to as the "pure face-on" 6P nanostripe. Instead of the formation of 6P bilayer nanostripes, further deposition of 6P results in the insertion of edge-on 6P molecules into the matrix of the "pure face-on" 6P nanostripes, where

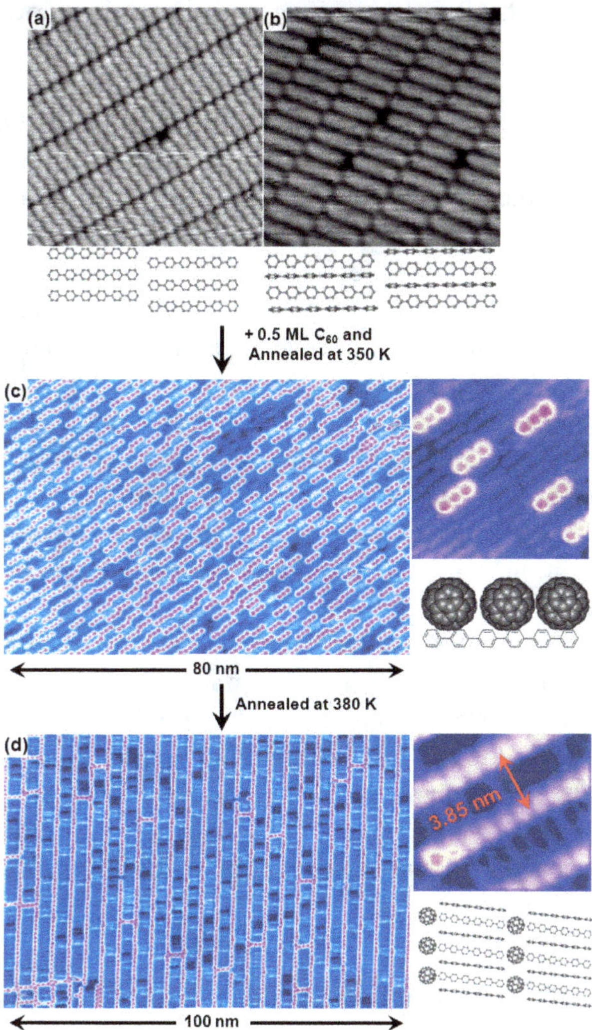

Fig. 4. $15 \times 15\,nm^2$ STM images of (a) "pure face-on" ($V_{tip} = +1.8$ V) and (b) "face-on + edge-on" ($V_{tip} = -3.3$ V) 6P monolayer nanostripe on Ag(111); the corresponding schematic drawings of molecular structure for both 6P nanostripes are shown below. (c) Left: STM image ($V_{tip} = +1.4$ V) of C_{60} triplet array or vertical C_{60}/6P nanojunction array formed after the deposition of 0.5 ML C_{60} on "face-on + edge-on" 6P layer followed by annealing at 350 K. Right: The corresponding detailed $15 \times 15\,nm^2$ STM image ($V_{tip} = +1.4$ V), showing the preferential trapping of every three C_{60} molecules on a single 6P molecule and the proposed molecular model. (d) Left: STM image ($V_{tip} = -2.0$ V) of C_{60} lateral C_{60}/6P nanojunction array formed after annealing the C_{60} triplet array at 380 K. Right: The corresponding detailed STM image ($V_{tip} = -2.0$ V) and the proposed molecular model for the regular superlattice of alternating C_{60} and 6P linear chains. (*Adapted from Ref. 79.*)

"edge-on" represents 6P molecules with their phenyl rings oriented nearly perpendicular to the substrate surface, as illustrated in Fig. 4(b). This results in the alternating arrangement of face-on and edge-on 6P molecules in this 6P nanostripe array, referred to as the "face-on + edge-on" 6P nanostripe. Such an alternating molecular arrangement is reminiscent of the herringbone structure commonly observed in 6P single-crystal solids, stabilized by the electrostatic force or the quadruple interaction between neighboring face-on and edge-on 6P molecules.[80,81]

Figure 4(c) shows the RT adsorption of 0.5 ML C_{60} on the "face-on + edge-on" 6P nanostripes with subsequent annealing at 350 K for 30 min. A fairly ordered 2D array of C_{60} triplets forms, originating from the preferential trapping of a C_{60} triplet atop each 6P molecule.[79] These triplets assemble along the underlying 6P nanostripe packing direction with a separation of about 3.3 nm. These well-ordered C_{60} triplet arrays atop the 6P nanostripes can serve as nanoscale organic donor/acceptor vertical nanojunction array devices such as rectifiers, with applications in molecular nanoelectronics.

After annealing at 380 K for 30 min, the C_{60} triplet array is transformed into a highly periodic C_{60} linear chain array [see Fig. 4(c)], involving the insertion of C_{60} linear chains into the "face-on + edge-on" 6P nanostripe array.[79] The periodicity or the interchain distance is measured to be 3.85 ± 0.02 nm. Such a highly periodic 2D arrangement of alternating linear chains of C_{60} (acceptor) and 6P (donor) makes this superlattice a prototype lateral donor/acceptor nanojunction array for potential applications in organic solar cells.

2.4. Hydrogen-Bonded 2D Binary Molecular Networks

Highly directional intermolecular hydrogen bonding can easily lead to the formation of short-range-ordered molecular assemblies or nanostructures such as dimers and trimers, with well-defined binding geometries and molecular conformations. The creation of long-range-ordered 2D molecular networks on surfaces usually requires strong molecule–substrate interfacial interactions via epitaxial or quasiepitaxial interlocking with single crystalline substrates,[82,83] or through substrate-mediated long-range

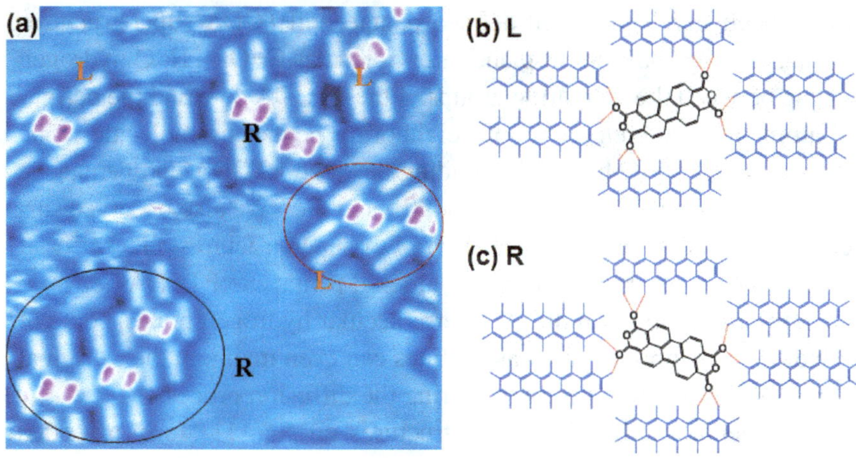

Fig. 5. (a) $14 \times 14\,\text{nm}^2$ STM image ($V_{\text{tip}} = 1.5\,\text{V}$) showing the coexistence of pentacene: PTCDA superstructures with L and R chirality, and their corresponding schematic drawings are shown in panels (b) and (c), where the dashed lines indicate the multiple in-plane intermolecular C=O$\cdots$H–C hydrogen bonding between the PTCDA core and six surrounding pentacene molecules. (*Adapted from Ref. 87.*)

intermolecular interactions.[84–86] Such strong molecule–substrate interfacial interactions can force molecules to assemble into well-ordered structures controlled by the surface periodicity. In this section, we demonstrate the formation of long-range-ordered 2D pentacene: 3,4,9,10-perylene-tetracarboxylic-dianhydride (PTCDA) supramolecular chiral networks on Ag(111), which are controlled by interlocking of both molecules with Ag(111), and further stabilized through the directional C=O$\cdots$H–C multiple intermolecular hydrogen bonding between PTCDA and pentacene.[87]

Figure 5(a) shows the $14 \times 14\,\text{nm}^2$ STM image of a disordered pentacene:PTCDA mixed phase on Ag(111) at 77 K. It is dominated by isolated but well-defined supramolecular nanostructures, made up of a PTCDA core surrounded by six pentacene molecules, where the center component with two bright and parallel stripes represents a PTCDA single molecule and the surrounding rodlike features are pentacene molecules.[87] These pentacene:PTCDA nanostructures possess two in-plane mirror arrangements, i.e. surface chiral superstructures. The schematic drawings of supramolecular arrangements of L and R assemblies are shown in Figs. 5(b) and

5(c), respectively. These well-defined pentacene:PTCDA superstructures are stabilized by multiple in-plane intermolecular C=O· · · H–C hydrogen bonding between the anhydride groups of the PTCDA core and the peripheral aromatic hydrogen atoms of neighboring pentacene molecules. The possible C=O· · · H–C hydrogen bonding between PTCDA and pentacene is highlighted by dashed lines in Figs. 5(b) and 5(c). We also observed surface diffusion of isolated pentacene:PTCDA superstructures on terraces at 77 K toward the existing short-range-ordered pentacene:PTCDA network. Due to the formation of multiple intermolecular hydrogen bonds, some of these diffused supramolecular assemblies can even maintain their original intermolecular bonding geometry during the diffusion process.

The pentacene:PTCDA superstructures on Ag(111) form extended ordered surface networks after the sequential deposition of 0.7 monolayer of pentacene and 0.2 monolayer of PTCDA at RT.[87] In contrast to the random mixing of L and R assemblies in the disordered pentacene:PTCDA mixed phase [Fig. 5(a)], ordered surface networks are exclusively constructed from L or R assemblies, and are referred to as L or R networks as shown in the molecularly resolved STM images in Figs. 6(a) and 6(b). The two networks are mirror domains with respect to each other. They adopt a p2 plane group symmetry with a rectangular unit cell of $a = 1.85 \pm 0.05$ nm, $b = 2.15 \pm 0.05$ nm and $\alpha = 90° \pm 3°$. Self-organization of molecules into ordered nanostructures in extended surface networks relies on the interplay of multiple interactions on different strengths and length scales.[1,11–14] Preferential in-plane orientation of both pentacene and PTCDA molecules on Ag(111) has been observed, i.e. the long molecular axis of pentacene orients approximately along the [1–10] direction of Ag(111), and that of PTCDA along the [1–21] direction. This implies a registry of adsorbed pentacene or PTCDA with Ag(111) and strong molecule–metal interfacial interactions. The Ag(111) substrate acts as a template constraining the lateral degrees of freedom of adsorbed PTCDA and pentacene molecules. Therefore, the formation of 2D pentacene:PTCDA supramolecular chiral networks on Ag(111) is controlled by the strong molecule–substrate interfacial interactions, and is further stabilized through multiple in-plane intermolecular H-bonds between the PTCDA core and the neighboring pentacene molecules, as shown by the schematic drawings in Fig. 6.

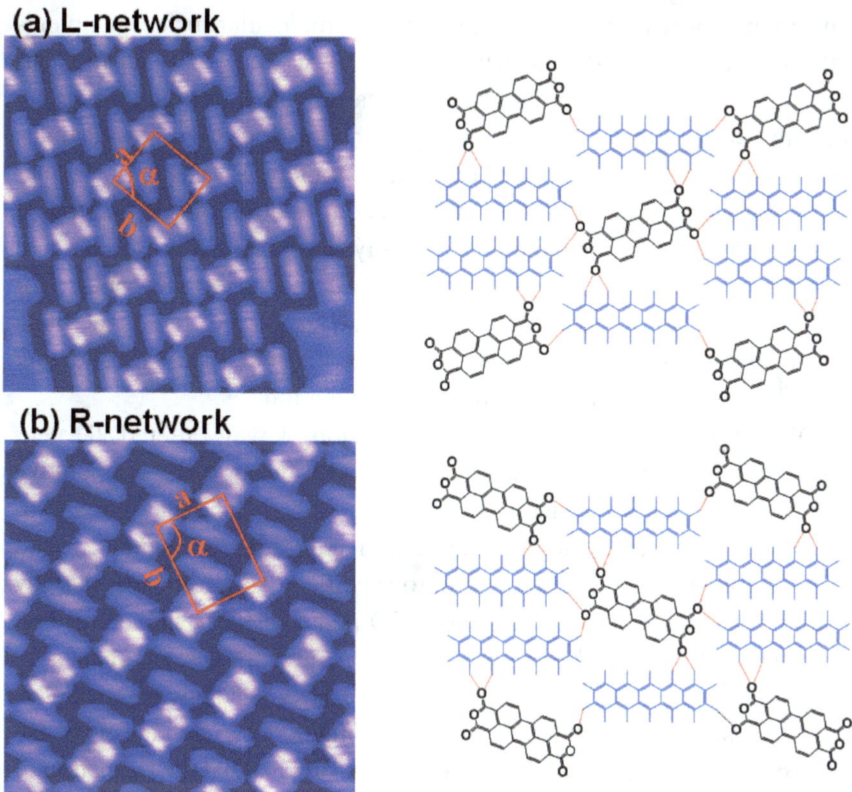

Fig. 6. STM images of a highly ordered pentacene:PTCDA 2D network with (a) L chirality (10×10 nm^2; $V_{tip} = 1.5$ V) and (b) R chirality (8×8 nm^2; $V_{tip} = 1.5$ V), and their corresponding schematic drawings are shown in the left panels, where the dashed lines indicate the multiple in-plane intermolecular C=O$\cdots$H–C hydrogen bonding between the PTCDA core and six surrounding pentacene molecules. The rectangles in the STM images highlight the unit cell with $a = 1.85 \pm 0.05$ nm, $b = 2.15 \pm 0.05$ nm and $\alpha = 90° \pm 3°$. (*Adapted from Ref. 87.*)

2.5. Conclusion and Perspectives

In conclusion, we have demonstrated the fabrication of various C_{60} nanostructure arrays using different molecular surface nanotemplates. On the robust C_{60} nanomesh template, the trapping of single C_{60} molecules into each nanocavity leads to the formation of regular C_{60} arrays, whose periodicity is determined by the underlying nanotemplate. In contrast to the conventional inorganic surface nanotemplates, molecular surface nanotemplates such as 6T or 6P nanostripes are sustained by weak intermolecular interactions, facilitating the structural rearrangement of the underlying molecular surface nanotemplate during the deposition of guest molecules by careful control of experimental conditions, thereby leading to the formation of tunable C_{60} nanostructure arrays. The formation of self-assembled molecular donor/acceptor nanojunction arrays has potential uses in organic photovoltaic cells to maximize the donor/acceptor interface contact for effective exciton dissociation into electrons and holes, as well as to facilitate the efficient transport of these electrons and holes into opposite collecting electrodes, minimizing electron–hole recombination.

We have also demonstrated the formation of well-ordered 2D pentacene:PTCDA supramolecular chiral networks through the cooperative arrangement of pentacene and PTCDA on Ag(111). Ag(111) acts as a template to restrict the lateral degrees of freedom of the adsorbed pentacene and PTCDA molecules, and locks them into specific adsorption sites, thereby favoring the formation of well-ordered 2D pentacene:PTCDA chiral networks. The structure of the pentacene:PTCDA network is strengthened by multiple in-plane intermolecular C=O$\cdots$H–C H-bonds between the anhydride groups of the PTCDA core and the aromatic hydrogen atoms of the neighboring pentacene molecules. Our results suggest that the formation of large-scale highly periodic 2D supramolecular assemblies often relies on a combination of specific and relatively strong molecule–substrate interactions with weak but directional intermolecular interactions. By rational design of molecules with molecular centers featuring key functionalities such as spins and peripheral atoms for intermolecular hydrogen bonding, the fabrication of long-range-ordered molecular nanostructure arrays with desired functionalities over the macroscopic area can be readily achieved.

Future research in this field will be directed toward the characterization of the unique electronic, electrical and optical properties of these self-assembled molecular nanostructure arrays for their applications in molecular electronics. The investigation of such molecular nanostructure arrays in solution or air is needed since these conditions are more favorable to commercial applications.

Acknowledgments

The authors acknowledge the support from the Singapore A*STAR grant R-398-000-036-305 and the MOE ARF grants R-144-000-196-112 and R-143-000-392-133.

References

1. J. V. Barth, G. Constantini and K. Kern, *Nature* **437**, 671 (2005).
2. C. Joachim and M. A. Ratner, *Proc. Natl. Acad. Sci. USA* **102**, 8801 (2005).
3. J. R. Heath and M. A. Ratner, *Phys. Today* **56**, 43 (2003).
4. Y. Selzer and D. L. Allara, *Annu. Rev. Phys. Chem.* **57**, 593 (2006).
5. S. C. Benjamin *et al. J. Phys. Condens. Matter* **18**, S867 (2006).
6. T. J. Park *et al.*, *Anal. Chem.* **78**, 7197 (2006).
7. N. L. Rosi and C. A. Mirkin, *Chem. Rev.* **105**, 1547 (2005).
8. A. S. Blawas and W. M. Reichert, *Biomaterials* **19**, 595 (1998).
9. J. C. Smith *et al.*, *Nano Lett.* **3**, 883 (2003).
10. M. A. Loi, E. Da Como, F. Dinelli, M. Murgia, R. Zamboni, F. Biscarini and M. Muccini, *Nat. Mater* **4**, 81 (2005).
11. F. Rosei, *J. Phys. Condens. Matter* **16**, S1373 (2004).
12. W. Chen and A. T. S. Wee, *J. Phys. D: Appl. Phys.* **40**, 6287 (2007).
13. F. Tao and S. L. Bernasek, *Chem. Rev.* **107**, 1408 (2007).
14. F. Rosei, M. Schunack, Y. Naitoh, P. Jiang, A. Gourdon, E. Laegsgaard, I. Stensgaard, C. Joachim and F. Besenbacher, *Prog. Surf. Sci.* **71**, 95 (2003).
15. L. F. Yuan *et al.*, *J. Am. Chem. Soc.* **125**, 169 (2003).
16. J. G. Hou *et al.*, *Nature* **409**, 304 (2001).
17. S. C. Li, J. F. Jia, R. F. Dou, Q. K. Xue, I. G. Batyrev and S. B. Zhang, *Phys. Rev. Lett.* **93**, 116103 (2004).
18. K. H. Wu, Y. Fujikawa, T. Nagao, Y. Hasegawa, K. S. Nakayama, Q. K. Xue, E. G. Wang, T. Briere, V. Kumar, Y. Kawazoe, S. B. Zhang and T. Sakurai, *Phys. Rev. Lett.* **91**, 126101 (2003).
19. J. L. Li, J. F. Jia, X. J. Liang, X. Liu, J. Z. Wang, Q. K. Xue, Z. Q. Li, J. S. Tse, Z. Y. Zhang and S. B. Zhang, *Phys. Rev. Lett.* **88**, 066101 (2002).
20. R. W. Li, J. H. G. Owen, S. Kusano and K. Miki, *Appl. Phys. Lett.* **89**, 073116 (2006).
21. M. A. K. Zilani, Y. Y. Sun, H. Xu, L. Liu, Y. P. Feng, X. S. Wang and A. T. S. Wee, *Phys. Rev. B* **72**, 193402 (2005).
22. I. Ošt'ádal, P. Kocán, P. Sobotík and J. Pudl, *Phys. Rev. Lett.* **95**, 146101 (2005).
23. Y. L. Wang, H. J. Gao, H. M. Guo, S. Wang and S. T. Pantelides, *Phys. Rev. Lett.* **94**, 106101 (2005).

24. Y. P. Zhang, L. Yang, Y. H. Lai, G. Q. Xu and X. S. Wang, *Surf. Sci.* **531**, L378 (2003).
25. J. F. Jia, X. Liu, J. Z. Wang, J. L. Li, X. S. Wang, Q. K. Xue, Z. Q. Li, Z. Y. Zhang and S. B. Zhang, *Phys. Rev. B* **66**, 165412 (2002).
26. V. G. Kotlyar, A. V. Zotov, A. A. Saranin, T. V. Kasyanova, M. A. Cherevik, I. V. Pisarenko and V. G. Lifshits, *Phys. Rev. B* **66**, 165401 (2002).
27. V. Repain, G. Baudot, H. Ellmer and S. Rousset, *Europhys. Lett.* **58**, 730 (2002).
28. C. Didiot, S. Pons, B. Kierren, Y. Fagot-Revurat and D. Malterre, *Surf. Sci.* **600**, 3917 (2006).
29. N. Weiss, T. Cren, M. Epple, S. Rusponi, G. Baudot, S. Rohart, A. Tejeda, V. Repain, S. Rousset, P. Ohresser, F. Scheurer, P. Bencok and H. Brune, *Phys. Rev. Lett.* **95**, 157204 (2005).
30. P. Gambardella, M. Blanc, H. Brune, K. Kuhnke and K. Kern, *Phys. Rev. B* **61**, 2254 (2000).
31. R. Otero, Y. Naitoh, F. Rosei, P. Jiang, P. Thostrup, A. Gourdon, E. Laegsgaard, I. Stensgaard, C. Joachim and F. Besenbacher, *Angew. Chem. Int. Ed.* **43**, 2092 (2004).
32. H. Brune, *Surf. Sci. Rep.* **31**, 121 (1998).
33. H. Brune, M. Giovannini, K. Bromann and K. Kern, *Nature* **394**, 451 (1998).
34. M. Corso, W. Auwärter, M. Muntwiler, A. Tamai, T. Greber and J. Osterwalder, *Science* **303**, 217 (2004).
35. H. Dil, J. Lobo-Checa, R. Laskowski, P. Blaha, S. Berner, J. Osterwalder and T. Greber, *Science* **319**, 1824 (2008).
36. J. A. Theobald, N. S. Oxtoby, M. A. Phillips, N. R. Champness and P. H. Beton, *Nature* **424**, 1029 (2003).
37. J. A. Theobald, N. S. Oxtoby, N. R. Champness, P. H. Beton and T. J. S. Dennis, *Langmuir* **21**, 2038 (2005).
38. J. Ma, B. L. Rogers, M. J. Humphry, D. J. Ring, G. Goretzki, N. R. Champness and P. H. Beton, *J. Phys. Chem. B* **110**, 12207 (2006).
39. L. M. A. Perdigão, E. W. Perkins, J. Ma, P. A. Staniec, B. L. Rogers, N. R. Champness and P. H. Beton, *J. Phys. Chem. B* **110**, 12539 (2006).
40. J. C. Swarbrick, B. L. Rogers, N. R. Champness and P. H. Beton, *J. Phys. Chem. B* **110**, 6110 (2006).
41. L. M. A. Perdigão, N. R. Champness and P. H. Beton, *Chem. Comm.* **538** (2006).
42. J. Weckesser, A. De Vita, J. V. Barth, C. Cai and K. Kern, *Phys. Rev. Lett.* **87**, 096101 (2001).
43. J. V. Barth, J. Weckesser, C. Cai, P. Günter, L. Bürgi, O. Jeandupeux and K. Kern, *Angew. Chem. Int. Ed.* **39**, 1230 (2000).
44. J. V. Barth, J. Weckesser, G. Trimarchi, M. Vladimirova, A. De Vita, C. Cai, H. Brune, P. Günter and K. Kern, *J. Am. Chem. Soc.* **124**, 7991 (2002).
45. A. Dmitriev, N. Lin, J. Weckesser, J. V. Barth and K. Kern, *J. Phys. Chem. B* **106**, 6907 (2002).
46. S. Griessl, M. Lackinger, M. Edelwirth, M. Hietschold and W. M. Heckl, *Single Mol.* **3**, 25 (2002).

47. K. G. Nath, O. Ivasenko, J. A. Miwa, H. Dang, J. D. Wuest, A. Nanci, D. F. Perepichka and F. Rosei, *J. Am. Chem. Soc.* **128**, 4212 (2006).
48. G. Pawin, K. L. Wong, K. Y. Kwon and L. Bartels, *Science* **313**, 961 (2006).
49. T. Huang, Z. Hu, A. Zhao, H. Wang, B. Wang, J. L. Yang and J. G. Hou, *J. Am. Chem. Soc.* **129**, 3857 (2007).
50. J. Schnadt *et al.*, *Phys. Rev. Lett.* **100**, 046103 (2008).
51. K. Tahara *et al.*, *J. Am. Chem. Soc.* **130**, 6666 (2008).
52. S. B. Lei *et al.*, *J. Am. Chem. Soc.* **130**, 7119 (2008).
53. S. B. Lei *et al.*, *Angew. Chem. Int. Ed.* **47**, 2964 (2008).
54. R. Otero, M. Lukas, R. E. A. Kelly, W. Xu, E. Lagsgaard, I. Stensgaard, L. N. Kantorovich and F. Besenbacher, *Science* **319**, 312 (2008).
55. N. Lin, A. Dmitriev, J. Weckesser, J. V. Barth and K. Kern, *Angew. Chem. Int. Ed.* **41**, 4779 (2002).
56. M. A. Lingenfelder, H. Spillmann, A. Dmitriev, S. Stepanow, N. Lin, J. V. Barth and K. Kern, *Chem. Eur. J.* **10**, 1913 (2004).
57. S. Stepanow, N. Lin, J. V. Barth and K. Kern, *Chem. Comm.* 2153 (2006).
58. S. Stepanow, N. Lin, J. V. Barth and K. Kern, *J. Phys. Chem. B* **110**, 23472 (2006).
59. N. Lin, S. Stepanow, F. Vidal, K. Kern, M. S. Alam, S. Stromsdorfer, V. Dremov, P. Muller, A. Landa and M. Ruben, *Dalton Trans.* 2794 (2006).
60. N. Lin, S. Stepanow, F. Vidal, J. V. Barth and K. Kern, *Chem. Comm.* 1681 (2005).
61. H. Spillmann, A. Dmitriev, N. Lin, P. Messina, J. V. Barth and K. Kern, *J. Am. Chem. Soc.* **125**, 10725 (2003).
62. S. Stepanow, M. Lingenfelder, A. Dmitriev, H. Spillmann, E. Delvigne, X. Deng C. Cai, J. V. Barth and K. Kern, *Nat. Mater* **3**, 229 (2004).
63. S. Stepanow, N. Lin and J. V. Barth, *J. Phys. Condens. Matter* **20**, 184002 (2008).
64. U. Schickum *et al.*, *Nano Letter.* **7**, 3813 (2008).
65. D. Kuehne *et al.*, *J. Am. Chem. Soc.* **131**, 3881 (2009).
66. P. Gambardella *et al.*, *Nat. Mater* **8**, 189 (2009).
67. H. Spillmann, A. Kiebele, M. Stöhr, T. A. Jung, D. Bonifazi, F. Y. Cheng and F. Diederich, *Adv. Mater* **18**, 275 (2006).
68. H. L. Zhang, W. Chen, H. Huang, L. Chen and A. T. S. Wee, *J. Am. Chem. Soc.* **130**, 2720 (2008).
69. Q. Chen and N. V. Richardson, *Nat. Mater* **2**, 324 (2003).
70. M. Bohringer, K. Morgenstern, W. D. Schneider and R. Berndt, *Angew. Chem. Int. Ed.* **38**, 821 (1999).
71. M. Bohringer, K. Morgenstern, W. D. Schneider, R. Berndt, F. Mauri, A. de Vita and R. Car, *Phys. Rev. Lett.* **83**, 324 (1999).
72. M. O. Lorenzo, C. J. Baddeley, C. Muryn and R. Raval, *Nature* **404**, 376 (2000).
73. W. Chen, H. Huang, S. Chen, L. Chen, H. L. Zhang, X. Y. Gao and A. T. S. Wee, *Appl. Phys. Lett.* **91**, 114102 (2007).
74. W. Chen, H. L. Zhang, H. Huang, L. Chen and A. T. S. Wee, *ACS Nano* **2**, 693 (2008).
75. H. Huang, W. Chen and A. T. S. Wee, *J. Phys. Chem. C* **112**, 14913 (2008).

76. H. L. Zhang, W. Chen, L. Chen, H. Huang, X. S. Wang, J. Yuhara and A. T. S. Wee, *Small* **3**, 2015 (2007).
77. L. Chen, W. Chen, H. Huang, H. L. Zhang, J. Yuhara and A. T. S. Wee, *Adv. Mater.* **20**, 484 (2008).
78. H. Huang, W. Chen, L. Chen, H. L. Zhang, X. S. Wang, S. N. Bao and A. T. S. Wee, *Appl. Phys. Lett.* **92**, 023105 (2008).
79. W. Chen, H. L. Zhang, H. Huang, L. Chen and A. T. S. Wee, *Appl. Phys. Lett.* **92**, 193301 (2008).
80. G. Koller, S. Berkebile, M. Oehzelt, P. Puschnig, C. Ambrosch-Draxl, F. P. Netzer and M. G. Ramsey, *Science* **317**, 351, (2007).
81. W. Chen, H. Huang and A. T. S. Wee, *Chem. Comm.* 4276 (2008).
82. F. Vonau, D. Suhr, D. Aubel, L. Bouteiller, G. Reiter and L. Simon, *Phys. Rev. Lett.* **94**, 066103 (2005).
83. F. Vonau, D. Aubel, L. Bouteiller, G. Reiter and L. Simon, *Phys. Rev. Lett.* **99**, 086103 (2007).
84. S. Lukas, G. Witte and Ch. Wöll, *Phys. Rev. Lett.* **88**, 028301 (2002).
85. T. Yokoyama, T. Takahashi, K. Shinozaki and M. Okamoto, *Phys. Rev. Lett.* **98**, 206102 (2007).
86. I. Fernandez-Torrente, S. Monturet, K. J. Franke, J. Fraxedas, N. Lorente and J. I. Pascual, *Phys. Rev. Lett.* **99**, 176103 (2007).
87. W. Chen, H. Li, H. Huang, Y. X. Fu, H. L. Zhang, J. Ma and A. T. S. Wee, *J. Am. Chem. Soc.* **130**, 12285 (2008).

Wei Chen is an Assistant Professor in the Department of Chemistry and the Department of Physics at the National University of Singapore (NUS). He received his Bachelor's degree in Chemistry from Nanjing University (China) in 2001 and his PhD from the NUS' Department of Chemistry in 2004, supervised by Prof. Loh Kian Ping and Prof. Andrew T. S. Wee. From 2004 to 2006, he worked as a postdoctoral researcher in the surface science group in the Department of Physics. In 2006, he was awarded the Lee Kuan Yew postdoctoral research fellowship. His research focuses on the molecular self-assembly on surfaces and surface nanotemplates, molecular and organic electronics–related interface problems, and graphene-related materials and devices. He has published over 60 papers in peer-reviewed journals on these topics, including 4 invited review articles.

Andrew T. S. Wee is a Professor of Physics and currently Dean of the Faculty of Science at the National University of Singapore (NUS). He graduated with a BA (Hons) in Physics (1984) from the University of Cambridge, and a DPhil (1990) from the University of Oxford on a Rhodes Scholarship. He is also Director of the Surface Science Laboratory. He has published over 300 papers in internationally refereed journals in the field of surface and nanoscale science, and is an editor of several academic journals, including *Applied Physics Letters, Surface and Interface Analysis, International Journal of Nanoscience* and *Surface Review and Letters*.

Ballistic Electron Emission Microscopy on Hybrid Metal/Organic/Semiconductor Interfaces

Chapter 3

Cedric Troadec*,† and Kuan Eng Johnson Goh‡

Abstract

Although electronic devices based on organic materials are beginning to hit the consumer market, there remain fundamental questions such as the band structure coupling between metal electrodes and the organic active layer, the nature of Fermi-level pinning, and the role of interface dipoles at such heterojunctions. Answering these questions would be vital for the perpetuation of organic electronics. This chapter will demonstrate the unique capability of scanning tunneling microscopy to interrogate such device interfaces with nanoscale resolution using a three-terminal measurement that employs hot electrons: ballistic electron emission microscopy (BEEM). By revealing the local nanoscale distribution of key parameters such as the interface energy barrier and interface transmission, BEEM offers opportunities for answering questions critical to the success of organic electronics. To that end, this chapter will highlight the recent advances of the BEEM technique for studying metal/organic/semiconductor hybrid systems.

3.1. Introduction

Industry adoption of organic electronics and the copious funding directed at their continued development have been fueled by the potential for cheaper

*Institute of Materials Research and Engineering, Agency for Science, Technology and Research (A*STAR), 3 Research Link, Singapore 117602.
†Email: cedric-t@imre.a-star.edu.sg
‡Email: gohj@imre.a-star.edu.sg

manufacturing and more diverse functionalities such as brighter and lower-power displays, printable electronics and flexible substrates.[1,2] At the same time, the feasibility of tailor-made molecules with specific characteristics opens up a whole world of possibilities, e.g. wearable electronics,[3] low-cost lighting on flexible substrates[4,5] and flexible solar cells,[6] which challenges the prevailing dominance of inorganic devices. It is little wonder therefore that the field of organic electronics is developing rapidly and techniques, usually associated with conventional semiconductors are now being applied to this field. However, the complexities introduced by organic materials also imply that the analysis has to become more sophisticated in order to provide meaningful characterization and description of device behavior.

Here we focus attention on a characterization technique known as ballistic electron emission microscopy (BEEM),[7,8] a scanning probe microscopy (SPM)–based technique invented in the late 1980s to study the electronic properties of metal/semiconductor interfaces at the nanoscale level. The purpose of this chapter is to highlight the advantage of BEEM for studying metal/organic/semiconductor hybrid systems which have recently garnered significant technological interest. A number of groups over the last five years have worked in this area using BEEM, such as those led by Karen Kavanagh (Simon Fraser University), Julia Hsu (Sandia National Lab) and Jurgen Smoliner (Vienna University of Technology). Others used more conventional methods like *IV*, *CV* and optical measurements, notably Antoine Kahn (Princeton University) and David Cahen (Weizmann Institute of Science). While their work is important, this chapter serves to highlight, in particular, the contribution of Singapore in the characterization of such hybrid systems, which began with the 2004 breakthrough of using BEEM to characterize the first energy barrier of a metal/organic interface,[9] followed by the continuing research toward the advancement of the technique for hybrid-type systems. Although parallel studies have been performed on other device types involving organics (such as metal/organic/metal or organic/metal/inorganic semiconductors[10]), these are beyond the scope of the chapter.

In Sec. 3.2, we will provide a brief description of the BEEM technique and the extent to which the technique has been applied in the context of metal/organic interfaces. Next, Sec. 3.3 will concentrate on the main thrust of this chapter, which is the use of BEEM on hybrid systems of metal/organic/inorganic semiconductors. We conclude in Sec. 3.4 by presenting an outlook on the use of BEEM in this emerging field.

3.2. General Introduction to Ballistic Electron Emission Microscopy

BEEM is a variation of scanning tunneling microscopy (STM) and was invented by Bell and Kaiser in 1988.[7,8] A detailed description of this technique is beyond the scope of this article. However, the interested reader can find that in a number of recent reviews.[11–14] Here, we shall provide only a brief description of the BEEM technique in order to lay down the fundamental concepts and terminology associated with the technique, sufficient to equip the reader for understanding its application to the organic devices studied in this chapter.

In BEEM, illustrated schematically in Fig. 1, a typical sample consists of a semiconductor overlaid on one side with a thin metal film (termed the "base") and having an ohmic contact on the opposite side. The base is grounded, and a bias voltage (V_{bias}) applied at the STM probe tip injects carriers toward the metal/semiconductor interface. As in conventional STM, the tip–sample separation is controlled by a feedback circuit that maintains a constant tunneling current (I_T). At energies (eV_{bias}) above the metal base Fermi energy, the hot carriers propagate ballistically through the metal, before impinging on the interface. If the energy of the carriers exceeds

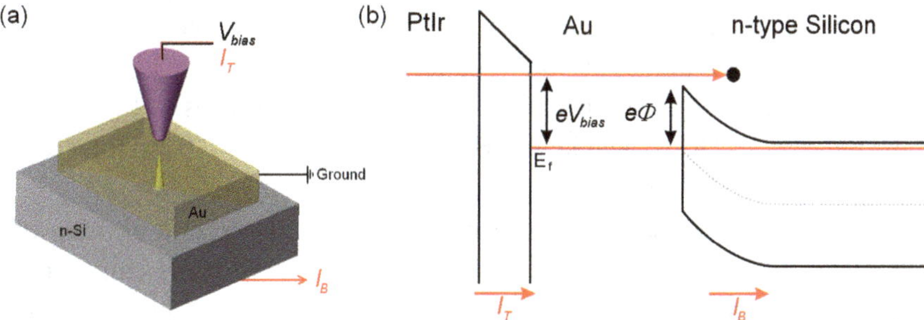

Fig. 1. (a) Schematic of a BEEM setup with an STM tip above a metal film (base) injecting carriers toward the metal/semiconductor interface. (b) Corresponding band diagram for electron injection at the interface.

the Schottky/injection barrier (Φ), they propagate into the semiconductor and can be collected via the ohmic contact. In BEEM measurements, the tunneling current is typically of the order of nA, whereas the BEEM current (I_B) is of the order of pA, depending on factors such as the transmission of the interface and the amount and type of scattering which occurs at the interface. We note that the name of the BEEM technique implies injection of electrons into the system, but injection of holes is also possible and this is sometimes termed "ballistic hole emission microscopy" (BHEM). In this chapter, references to "BEEM" will imply electron injection from the tip, unless otherwise stated.

In imaging mode, a fixed tip bias is maintained while the tip height (in constant I_T mode) and the BEEM current are concurrently tracked as a function of the tip position, allowing the STM topography of the metal surface and the BEEM image of the interface to be obtained simultaneously. In spectroscopy mode, the BEEM current is measured as a function of the tip bias at a fixed location, and a constant tunnel current is maintained to produce an $I_B - V_{\text{bias}}$ spectrum, also referred to as the ballistic electron emission spectroscopy (BEES) spectrum. Typical BEES refers to a (forward) voltage bias injecting majority carriers into the interface, whereas reverse BEES (RBEES) corresponds to injection of minority carriers toward the interface by applying a reverse voltage bias on the STM tip.

Many conventional techniques used for compositional (e.g. X-ray, ultraviolet photoelectron and Auger spectroscopy) and electrical (e.g. current–voltage and capacitance–voltage measurements) characterization of interfaces usually probe an area (frequently called the "spot size") in the range of at least a few μm^2. In contrast, BEEM, by virtue of being an STM-based technique, has the unique advantage of a very high spatial resolution, allowing local electronic properties of interfaces to be probed at the nanoscale. The typical lateral BEEM resolution is 1–10 nm for a thin metal film of about 10 nm.[15,16]

While BEEM imaging can provide a nanometer resolution map of ballistic electron transmission at the buried interface, BEES is commonly employed to provide insight into interface band structures by analyzing the threshold voltage(s) and transmission attenuation observed in the BEES spectrum. Figure 2 summarizes schematically the spectra associated with various BEES modes.

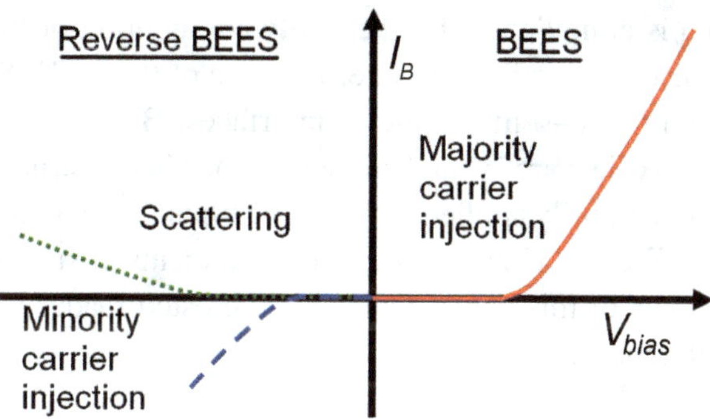

Fig. 2. Schematic of the different spectroscopic measurements available in BEEM. Majority and minority carriers refer to the semiconducting substrate which is part of the collector for the BEEM current.

The majority of BEEM studies have been carried out with majority carrier injection, and fitting of the BEES spectrum is well described by the Bell–Kaiser formalism near the threshold:

$$\frac{I_B}{I_T} = \frac{R(eV_{\text{bias}} - \Phi)^2}{eV_{\text{bias}}}, \tag{1}$$

where Φ (eV) is the local Schottky barrier of the base/semiconductor interface and R ($1/eV$) is a measure of the local transmission of hot majority carriers through the interface. To compensate[7,10,17] for the slight change in the tip–sample distance separation during BEES, the first-order normalization factor $1/eV_{\text{bias}}$ appears on the right-hand-side denominator of Eq. (1). The quadratic behavior is in good agreement with the experimental data near the threshold, although higher-order power laws can be used to describe the onset of the BEEM current in other cases.[11]

Under reverse bias (injecting minority carriers through the interface), the BEEM current is expected to have an opposite sign compared to the forward-biased BEEM current, since minority carriers are collected after passing through the interface. Indeed, this was observed in some reports.[18,19] However, most reported RBEES spectra show the same sign and similar threshold as the forward BEES spectra (dotted line in Fig. 2), which has been attributed to inelastic decay via electron–hole pair excitation (Auger-like process).[18,20] In addition, there is a possible contribution from STM photoinduced current (not BEEM-related) which is not quenched even for base thicknesses well beyond the carrier mean free path.[21] The analysis of RBEES data is complicated by the likely competition of these effects, which should be considered in the interpretation of the RBEES spectra.

Apart from metal–semiconductor interfaces, BEEM is widely used for the study of oxides[11,22–24] and semiconductor heterostructures.[25,26] It has also been successfully applied to magnetic films[27,28] and tips[29] to study spin-polarized effects.[30] However, detailed treatments of such systems are not the main focus of this chapter and the interested reader is referred to the relevant references.

3.3. BEEM in Hybrid Metal/Organic/Semiconductor Devices

The excitement about hybrid metal/organic/semiconductor research is fueled by the possibility of tailoring band alignments at device interfaces afforded by the wide-ranging functionality of molecules and the flexibility of their synthesis.[31–33] Although conventional spectroscopy techniques (as mentioned in Sec. 3.2) have been used to characterize hybrid molecular diodes, the information obtained is limited to micron spatial resolution and to averages of the parameters measured. The nanometer scale resolution of BEEM offers the unique advantage of studying the effects of local changes in key parameters like barrier height and transmission, and hence provides detailed insight into the device performance otherwise not attainable with conventional techniques.

Hybrid metal/semiconductor diodes can be divided into two categories, depending on how the molecules are attached to the semiconductor: chemisorbed or physisorbed.

3.3.1. *Chemisorbed molecule*

Chemisorbed molecules typically refer to molecules which share a strong bond with the substrate. For example, maleic anhydride ($C_4H_2O_3$) can be attached as a monolayer to the Si(111) surface via a covalent bond between one of its carbon atoms and a silicon (Si) atom from the substrate.[34,35] Its carboxyl functional groups can be modified into other functional groups to allow the chemisorbed monolayer to be functionally modified. We have performed BEEM studies on such a self-assembled monolayer (SAM) sandwiched between a 15 nm-thick thermally evaporated gold (Au) film (diameter 1 mm) and a low n-doped Si(111) substrate.[36] Performing conventional I–V measurement on this sample revealed a barrier height of ~0.80 eV, which is indistinguishable from that of a Au/n-Si diode. In contrast, by performing BEES over an area of $500 \times 500\,\text{nm}^2$, we found a distribution of barrier heights, as shown in Fig. 3(a). Figure 3(b) is an example of the BEES spectrum taken on this Au/maleic anhydride/n-Si(111) sample along with the conformation of the molecule on the substrate.

From Fig. 3(a), we can see that the barrier height distribution has a peak value of about 1 eV, which is higher than the Au/n-Si(111) barrier height. This suggests that the maleic anhydride monolayer in this local region is likely to be better-formed, with less pinholes of Au in contact with Si. Here BEEM shows the clear advantage of being able to discriminate between interfaces that would otherwise look similar using I–V characterization.

Other groups have also studied the impact of a SAM inserted between an Au layer and a Si substrate. In the report by Kuikka *et al.*, the BEEM spectra showed no evidence of any barrier height change associated with the SAM.[37] This is in contrast to hybrid Au/SAM/n-GaAs devices (typical barrier height ~0.9 eV), in which sandwiched carboxylic acid monolayers[38] or alkanethiol monolayers[39,40] caused the barrier detected by BEEM to

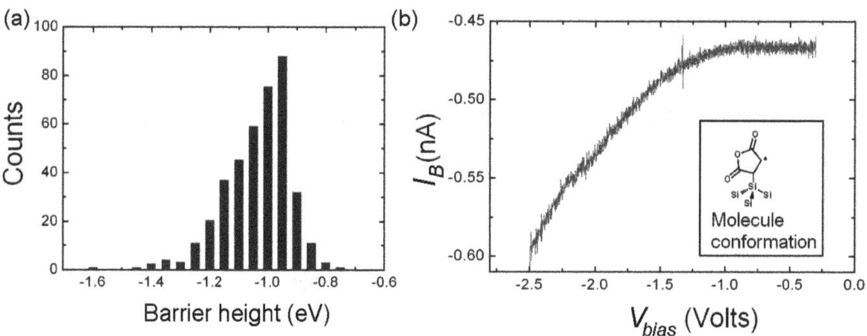

Fig. 3. (a) Distribution of barrier heights from about 400 BEES spectra taken over a 500 nm × 500 nm area. (b) BEES spectrum taken on a Au/maleic anhydride/n-Si(111) device at $I_T = 5$ nA, showing a barrier of ~0.87 eV. The inset indicates the molecule conformation on the Si substrate.

increase up to as much as 1.4 eV. As for the interface transmission, the inclusion of a chemisorbed organic layer on Si or GaAs invariably resulted in a decrease in the BEEM current as compared to the bare metal/semiconductor diode.[37,41]

3.3.2. *Physisorbed molecule*

BEEM measurements on Au-covered molecular film of titanyl phthalocyanine (TiOPc)[42] weakly bound on n-GaAs have been reported recently by Özcan *et al.* The barrier height at the Au/TiOPc interface was found to increase approximately linearly from 1.2 eV to 1.5 eV as the temperature was reduced from room temperature to 10 K. The report also presents a method to extract the attenuation length of the hot electrons in the buried TiOPC film[43] (found to be about 4 Å at energies above 1.25 eV and room temperature). The same group also reported BEEM measurements at different temperatures on a Au/hexa-*peri*-hexabenzocoronene (HBC) molecule interface[44] on GaAs. A barrier height of 1.3 eV was reported for the Au/HBC interface. Other physisorbed molecules deposited by evaporation or spin coating, but in a metal/molecule/metal configuration, have also been reported.[9,45]

In the following section, we will present a case study performed on a physisorbed organic layer deposited on Si, covered by Au. We shall use this example to motivate what we call a dual-parameter BEES analysis, which we believe is an important step toward visualizing the effect of inhomogeneous interfaces in terms of barrier height and transmission.

3.4. BEEM on Hybrid Au/Pentacene/*n*-Si Interfaces

Here we will use BEEM to study a Au/pentacene/Si interface. Pentacene is a well-studied organic semiconductor, in part motivated by its relatively high electron mobility and hence its potential for low-cost organic electronics. Pentacene thin film growth on the hydrogen-passivated n-Si(111) surface is known to produce a mixed phase film comprising crystalline islands surrounded by amorphous phase molecules, often with "pinholes", or areas without any molecules.[46] This creates opportunities for the study of metal-organic contact issues such as incomplete organic coverage or metal diffusion through the organic layer. All the fabrication and analytical details are available in recent publications.[47,48] For clarity, we present here a brief summary of the sample preparation and measurement setup.

The Au/pentacene/Si diodes were fabricated on n-type Si(111) substrates (1–10 Ω cm) of $6 \times 6\ \text{mm}^2$. The substrate backside was ion-implanted to provide an ohmic contact for the collection of the BEEM current. After standard chemical cleaning of the Si substrate, a brief dip in aqueous NH_4F ensured proper hydrogen passivation of the silicon surface. Pentacene was then sublimated onto a predetermined portion of the chip by shadow masking to obtain a nominal layer thickness of 1.5 nm. For such a thin film of pentacene, previous work by Shimada *et al.* indicates that incomplete pentacene coverage is expected.[46] Finally, a 15 nm Au film was deposited on top to form two types of circular diodes: Au/n-Si(111) and Au/pentacene/n-Si(111) (see Fig. 4). This allows direct comparison between the two systems.

The measurements (BEEM and I–V) were performed in a home-built BEEM–STM system comprising a modified Molecular Imaging STM head, a XYZ Attocube stage for sample placement (up to 6 mm in each axis), a Kleindiek micromanipulator to gently ground the base with a 50 μm

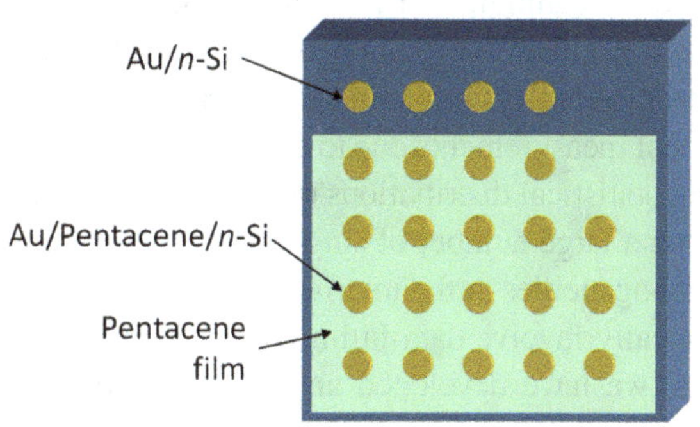

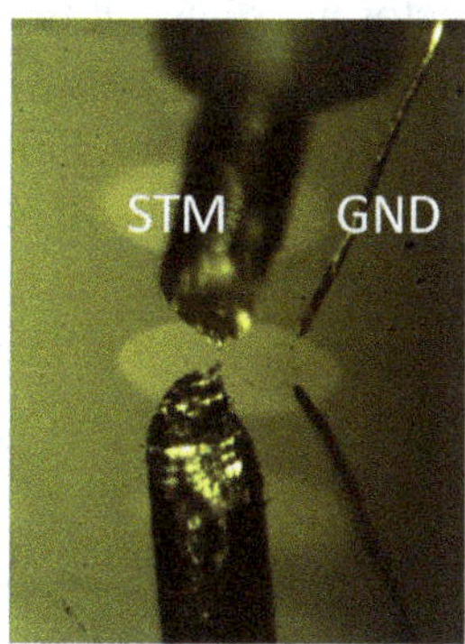

Fig. 4. Left: Schematic of the sample layout with the two types of diodes made on the n-Si(111) substrate (Au/n-Si and Au/pentacene/n-Si). Right: Snapshot of the actual device under BEEM measurement. The thicker wire is the PtIr STM tip. The finer wire next to it is a 50 μm Au wire used to ground the metal base.

Au wire and a Nanonis controller. The system can be operated in air or low vacuum at ambient temperature. The back ohmic contact is directly touching a conducting sample stage which acts as the common collector for the BEEM current (connected to a FEMTO amplifier), or the common ground for *I-V* measurements. Cut PtIr tips were used for the BEEM–STM measurements.

3.4.1. *Density plots of barrier height and transmission*

Spectroscopic BEEM measurements are commonly used to evaluate barrier heights or threshold energies associated with the buried interface. The transmission has been less of a focus, mostly used in the determination of the carrier mean free path in different base materials at fixed voltages.[30,49–52] In this subsection, we will demonstrate that by including the transmission factor with the barrier height in a dual-parameter statistical analysis, new possibilities arise for understanding the Au/pentacene/*n*-Si system.

In systems which yield low collector currents in BEEM, numerous spectra taken over representative regions of interest are often averaged together to increase the signal-to-noise ratio. While this averaging is useful in systems which are quite homogeneous, like metal/inorganic semiconductor interfaces, it becomes less meaningful for inhomogeneous systems where an intermediate layer (e.g. organic) is present between the metal and the semiconductor. For inhomogeneous systems, such an averaging procedure loses the benefit of the local measurements due to BEEM. In contrast, we shall demonstrate here that statistical distributions of the barrier height and/or transmission factor from a large number of single spectra are more useful for understanding inhomogeneities at the interface.

Spectroscopic studies usually involve data-fitting a large number of spectra. To facilitate analysis, we have developed an in-house software called the Automated BEEM Spectra Analyser (ABSA)[53] in order to automate the data-fitting process. Briefly, the software extracts the barrier height Φ and the transmission factor R by fitting the BEEM spectra with Eq. (1). This gives two distributions for each grid of spectra: a barrier height distribution and a transmission distribution. In Fig. 5 the respective distributions

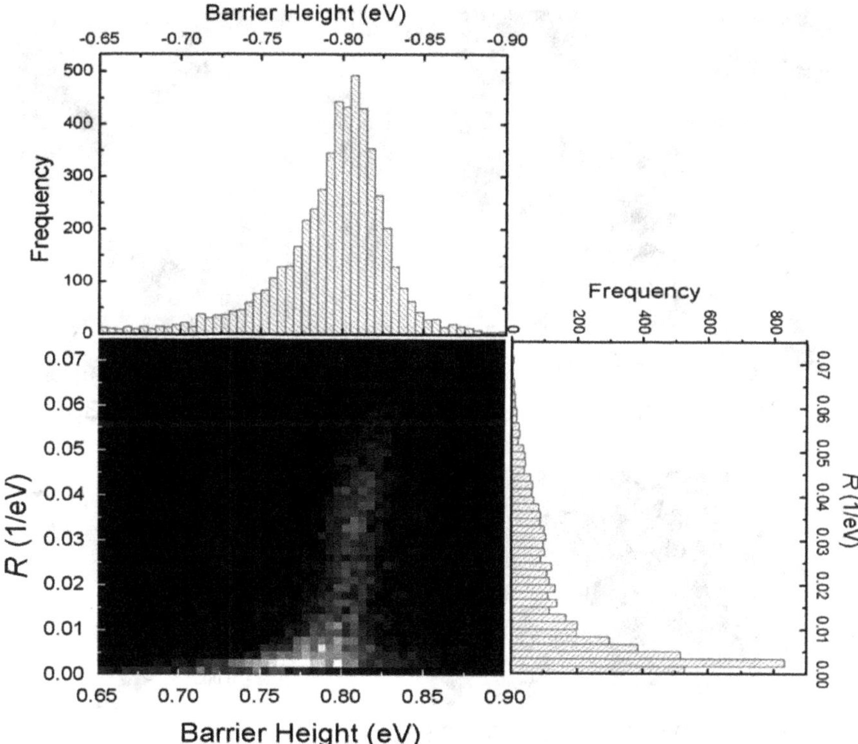

Fig. 5. The two outer graphs are the distribution of the Schottky barrier height and transmission factor for a sample of about 4000 randomly taken spectra for one Au/pentacene/*n*-Si(111) device. These two plots can be combined in one graph, in the bottom-left, which represents the density distribution of barrier height versus transmission factor.

from a few grids for a Au/pentacene/*n*-Si interface with the corresponding dual-parameter density representation are shown.

The density distribution of barrier height versus transmission factor plays a pivotal role in the understanding of the inhomogeneous system. First, the data with the experimental R value of $\sim 0.05\ \mathrm{eV}^{-1}$ may be attributed to the Au/*n*-Si interface [see Fig. 6(c), where the ellipse represents the expected position of the Au/*n*-Si distribution[47]]. Then, the decrease in transmission due to a thin tunneling barrier[54,55] (represented here by the pentacene layer) is estimated, taking into account the different types of morphology expected for a 1.5 nm film of pentacene[46] with lying and standing molecules. These

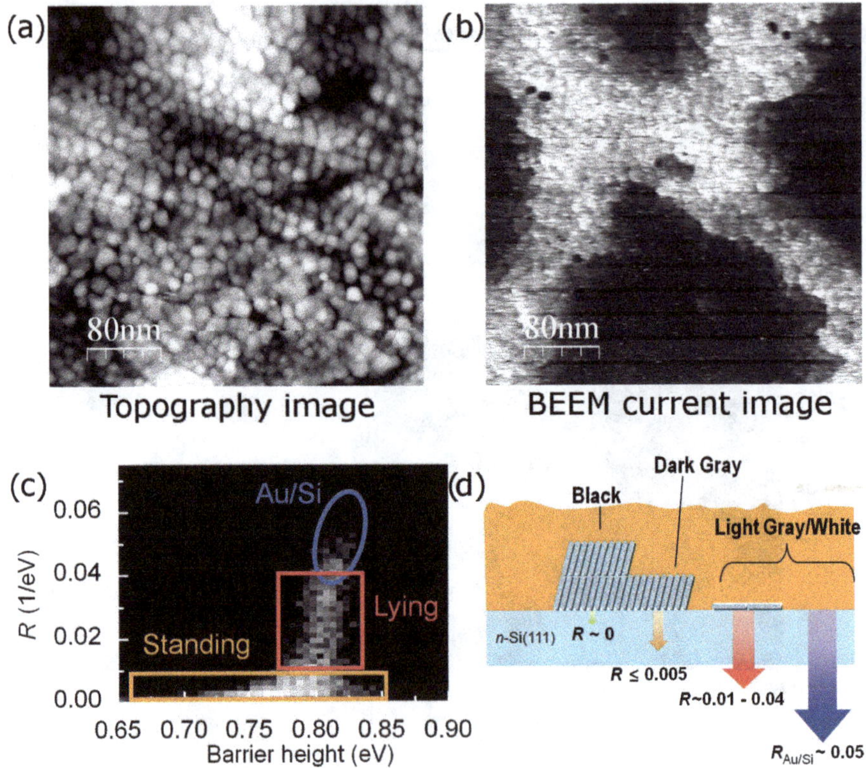

Fig. 6. (a) STM topography of Au/pentacene/n-Si taken at $V_{bias} = -1.5$ V and $I_T = 100$ pA. The full grayscale represents 11.7 nm. (b) Corresponding BEEM image, with the full grayscale representing 2.7 pA. (c) Density distribution of the Au/pentacene/n-Si device with the corresponding orientation of the pentacene molecules from the calculated transmission factors in (d), taking the experimental transmission factor of Au/n-Si as $R \sim 0.05\,\text{eV}^{-1}$.

calculated transmissions correlate well with the experimental transmission factors observed [see Fig. 6(d)], suggesting that the data points come from regions with lying and standing pentacene molecules. An example of STM topography and the corresponding BEEM image for the Au/pentacene/n-Si system is shown in Figs. 6(a) and 6(b), respectively. The topography image reveals the typical gold grain morphology. The presence or absence of buried pentacene molecules is revealed in the BEEM image as explained above, whereas the STM topography does not exhibit significant correlation with the interface imaged by BEEM.

By performing BEES and analyzing the results in dual-parameter space, we demonstrated the clear effect of the pentacene film and were able to correlate local transmission characteristics observed at the interface with the known morphology of thin film pentacene grown on Si. For detailed explanations, the reader is invited to read the full papers.[47,48]

The dual-parameter analysis that we have performed here should be useful in general for achieving a more detailed understanding of interface modifications by organics, defects or contaminants. We believe that this representation has clear advantages over simple barrier height or transmission factor distributions.

3.5. Conclusions and Outlook

Although the BEEM technique is over 20 years old, its application to new systems, especially metal/organic/semiconductor hybrid electronics, continues to stir interest. As a key technique capable of probing the electronic properties of interfaces at the nanoscale, it plays an increasingly important role complementary to more traditional surface-sensitive tools like XPS, UPS and XRD, which have only micrometer spatial resolution, and nonlocal transport analyses like *IV* and *CV*. As we have shown here, the transmission factor, a parameter often overlooked except in the measurement of hot electron attenuation length in metal, can be a useful parameter for mapping the morphology of underlying films.

Thus far, the theoretical efforts in BEEM have been concentrated on metal/inorganic semiconductor interfaces. The formalism pioneered by Bell and Kaiser[7,8] in 1988 remains the workhorse for most experimental and theoretical studies. It relies on the main assumption that the lateral momentum (*k* vector) at the interface is conserved. This is simple enough in the case of metal and crystalline inorganic semiconductor interfaces, but for organics which lack a well-defined *k* vector, it remains unclear if this conservation of lateral momentum at the interface should hold, if at all. Future theoretical and experimental treatments of BEEM will need to encompass the organic properties (mobility, highest occupied and lowest unoccupied molecular orbitals, conformation, hopping transport, etc.) in order to model these new types of interfaces. So far, there has only been one report[56] on the modeling of BEES on individual buried molecules. Such studies will undoubtedly refine both our theoretical and practical understanding of the metal/molecule interface, and form the next important phase in the development of the BEEM technique.

Acknowledgments

We would like to acknowledge research funding from A*STAR and IMRE. We thank both past and present members of the BEEM group for their contributions to the development of the BEEM systems and experiments, and for valuable discussions: Amin Bannani, Hailang Qin, Ying Qi Lee, Arpun Nagaraja, Han Fei Lai, Linda Kunardi, Justin Wen Chien Song and Natarajan Chandrasekhar.

References

1. H. Yan, Z. H. Chen, Y. Zheng, C. Newman, J. R. Quinn, F. Dotz, M. Kastler and A. Facchetti, *Nature* **457**, 679 (2009).
2. R. A. Street, W. S. Wong, S. E. Ready, I. L. Chabinyc, A. C. Arias, S. Limb, A. Salleo and R. Lujan, *Mater. Today* **9**, 32 (2006).
3. Y. Qin, X. Wang and Z. L. Wang, *Nature* **451**, 809 (2008).
4. M. Berggren, D. Nilsson and N. D. Robinson, *Nat. Mater.* **6**, 3 (2007).
5. C.-Y. Cheng and F. C.-N. Hong, *Jpn. J. Appl. Phys.* **45**, 8915 (2006).
6. Z. Fan, H. Razavi, J.-W. Do, A. Moriwaki, O. Ergen, Y.-L. Chueh, P. W. Leu, J. C. Ho, T. Takahashi, L. A. Reichertz, S. Neale, K. Yu, M. Wu, J. W. Ager and A. Javey, *Nat. Mater.* **8**, 648 (2009).
7. L. D. Bell and W. J. Kaiser, *Phys. Rev. Lett.* **61**, 2368 (1988).
8. W. J. Kaiser and L. D. Bell, *Phys. Rev. Lett.* **60**, 1406 (1988).
9. C. Troadec, D. Jie, L. Kunardi, S. J. O'Shea and N. Chandrasekhar, *Nanotechnology* **15**, 1818 (2004).
10. A. Bannani, C. Bobisch and R Moeller, *Science* **315**, 1824 (2007).
11. M. Prietsch, *Phys. Rep.* **253**, 163 (1995).
12. J. Smoliner, D. Rackoczy and M. Kast, *Rep. Prog. Phys.* **67**, 1863 (2004).
13. P. L. de Andres, F. J. Garcia-Vidal, K. Reuter and F. Flores, *Prog. Surf. Sci.* **66**, 3 (2001).
14. W. Yi, A. J. Stollenwerk and V. Narayanamurti, *Surf. Sci. Rep.* **64**, 169 (2009).

15. A. M. Milliken, S. J. Manion, W. J. Kaiser, L. D. Bell and M. H. Hecht, *Phys. Rev. B* **46**, 12826 (1992).
16. A. Davies, J. G. Couillard and H. G. Craighead, *Appl. Phys. Lett.* **61**, 31 (1992).
17. R. Ludeke and M. Prietsch, *J. Vac. Sci. Technol. A* **9**, 885 (1991).
18. R. Ludeke, *J. Vac. Sci. Technol. A* **11**, 786 (1993).
19. A. Chahboun, R. Coratger, R. Pechou, F. Ajustron and J. Beauvillain, *Surf. Sci.* **462**, 61 (2000).
20. L. D. Bell, M. H. Hecht, W. J. Kaiser and L. C. Davis, *Phys. Rev. Lett.* **64**, 2679 (1990).
21. E. R. Heller and J. P. Pelz, *App. Phys. Lett.* **82**, 3919 (2003).
22. Y. Zheng, A. T. S. Wee, K. L. Pey, C. Troadec, S. J. O'Shea and N. Chandrasekhar, *Appl. Phys. Lett.* **90**, 142915 (2007).
23. W. Cai, S. E. Stone, J. P. Pelz, L. F. Edge and D. G. Schlom, *Appl. Phys. Lett.* **91**, 042901 (2007).
24. Y. Zheng, A. T. S. Wee, Y. C. Ong, K. L. Pey, C, Troadec, S. J. O'Shea and N. Chandrasekhar, *Appl. Phys. Lett.* **92**, 012914 (2008).
25. W. Yi, V. Narayanamurti, J. M. O. Zide, S. R. Bank and A. C. Gossard, *Phys. Rev. B* **75**, 115333 (2007).
26. C. Tivarus, J. P. Pelz, M. K. Hudait and S. A. Ringel, *Phys. Rev. Lett.* **94**, 206803 (2005).
27. W. H. Rippard and R. A. Buhrman, *Appl. Phys. Lett.* **75**, 1001 (1999).
28. T. Banerjee, E. Haq, M. H. Siekman, J. C. Lodder and R. Jansen, *Phys. Rev. Lett.* **94**, 027204 (2005).
29. A. J. Stollenwerk, M. R. Krause, J. J. Garramone, E. J. Spadafora and V. P. LaBella, *Phys. Rev. B* **76**, 195311 (2007).
30. A. Kaidatzis, S. Rohart, A. Thiaville and J. Miltat, *Phys. Rev. B* **78**, 174426 (2008).
31. J. G. Kushmerick, *Mater. Today* **7&8**, 26 (2005).
32. D. Cahen, A. Kahn and E. Umbach, *Mater. Today* **7&8**, 32 (2005).
33. J. W. P. Hsu, *Mater. Today* **7&8**, 42 (2005).
34. Y. Liu, S. Yamazaki and S. Izuhara, *J. Org. Chem.* **691**, 5809 (2006).
35. Y. Liu, S. Yamazaki, S. Yamabe and Y. Nakato, *J. Mater. Chem.* **15**, 4906 (2005).
36. Y. Q. Lee, private communication (2008).
37. M. A. Kuikka, W. Li, K. L. Kavanagh and H. Yu, *J. Phys. Chem. C* **112**, 9081 (2008).
38. H. Haick, J. P. Pelz, T. Ligonzo, M. Ambrico, D. Cahen, W. Cai, C. Marginean, C. Tivarus and R. T. Tung, *Phys. Stat. Sol. (a)* **203**, 3438 (2006).
39. W. Li, K. L. Kavanagh, A. A. Talin, W. M. Clift, C. M. Matzke and J. W. P. Hsu, *J. Appl. Phys.* **102**, 13703 (2007).
40. W. Li, K. L. Kavanagh, C. M. Matzke, A. A. Talin, F. Leonard, S. Faleev and J. W. P. Hsu, *J. Phys. Chem. B* **109**, 6252 (2005).
41. A. Ahktari-Zavareh, W. Li, K. L. Kavanagh, A. J. Trionfi, J. C. Jones, J. L. Reno, J. W. P. Hsu and A. A. Talin, *J. Vac. Sci. Technol. B* **26**, 1597 (2008).

42. S. Özcan, J. Smoliner, M. Andrews, and G. Strasser, T. Dienel, R. Franke and T. Fritz, *Appl. Phys. Lett.* **90**, 92107 (2007).
43. S. Özcan, J. Smoliner, A. M. Andrews, G. Strasser, T. Dienel, R. Franke and T. Fritz *Semicond. Sci. Technol.* **23**, 055008 (2008).
44. S. Özcan, J. Smoliner, T. Dienel and T. Fritz, *Appl. Phys. Lett.* **92**, 153309 (2008).
45. C. Troadec, L. Kunardi and N. Chandrasekhar, *Appl. Phys. Lett.* **86**, 072101 (2005).
46. T. Shimada, H. Nogawa, T. Hasegawa, R. Okada, H. Ichikawa, K. Ueno and K. Saiki, *Appl. Phys. Lett.* **87**, 061917 (2005).
47. K. E. J. Goh, A. Bannani and C. Troadec, *Nanotechnology* **19**, 445718 (2008).
48. J. C. W. Song, K. E. J. Goh, N. Chandrasekhar and C. Troadec, *Phys. Rev. B* **79**, 165313 (2009).
49. L. D. Bell, *Phys. Rev. Lett.* **77**, 3893 (1996).
50. R. P. Lu, B. A. Morgan, K. L. Kavanagh, C. J. Powell, P. J. Chen, F. G. Sherpa and W. F. Egelhoff Jr., *J. Vac. Sci. Technol. B* **18**, 2047 (2000).
51. A. J. Stollenwerk, M. R. Krause, J. J. Garramone, E. J. Spadafora and V. P. LaBella, *Phys. Rev. B* **76**, 195311 (2007).
52. T. Banerjee, J. C. Lodder and R. Jansen, *Phys. Rev. B* **76**, 140407(R) (2007).
53. http://digital.ni.com/worldwide/singapore.nsf/web/all/BB79D6400D4EFD0E862574D70033E97
54. A. Salomon, T. Boecking, C. K. Chan, F. Amy, O. Girshevitz, D. Cahen and A. Kahn, *Phys. Rev. Lett.* **95**, 266807 (2005).
55. S. M. Sze, *Physics of Semiconductor Devices*, 2nd ed. (Wiley, New York, 1981) pp. 73–75.
56. G. Kirczenow, *Phys. Rev. B* **75**, 045428 (2007).

Kuan Eng Johnson Goh is currently a Senior Research Engineer at the Institute of Materials Research and Engineering (Singapore) where he leads the efforts in using scanning probe microscopy for atomic-scale fabrication and material interface characterisation. He completed his PhD in Physics at the University of New South Wales in 2006 under an Endeavour International Postgraduate Research Scholarship awarded by the Australian government and three International Fellowships awarded by A*STAR (Singapore). His thesis dealt with the encapsulation of Si:P delta-doped devices fabricated by scanning tunnelling microscopy and the magnetotransport in such devices. His current research addresses key issues in the fabrication of organic photovoltaic devices using ballistic electron emission microscopy, the reliability of high-K dieletrics, and the possibility of atomic precision manufacturing.

Cedric Troadec completed his PhD in Physics at the Royal Holloway College, University of London in 2001. His thesis was on hybrid superconductor/ferromagnetic metallic nanostructures: fabrication and study of the proximity effect. After working for Thermo VG Semicon designing gas cells for molecular beam epitaxy, he joined the Institute of Materials Research and Engineering (Singapore) in 2003, where he presently works as a Senior Research Engineer, to develop ballistic electron emission microscopy for the characterization of metal/organic interfaces.

Force–Extension Behavior of Single Polymer Chains by AFM

Chapter 4

Marina I. Giannotti, Edit Kutnyánszky and G. Julius Vancso*

Abstract

Applications of atomic force microscopy (AFM)–based single molecule force spectroscopy (SMFS) to mechanically stretch macromolecules and quantitatively probe single chain force–extension behavior in different media are reviewed. The models most frequently used to describe single chain entropic and enthalpic elasticity are introduced. The use of these models to fit single chain force–extension behavior so as to obtain single chain parameters such as the Kuhn segment length and segment elasticity is critically analyzed. Single chain behavior of stimulus-responsive synthetic polymers receives particular attention. Such macromolecules can exhibit reversible changes of single chain elasticity parameters upon variation of external stimuli, which can be utilized to convert external energy into mechanical work. Conversion of electromagnetic (photonic), as well as electrochemical, energy into work by the corresponding single chain macromolecular motors is given an overview. The results demonstrate that in addition to novel fundamental insights into the single chain behavior by AFM-SMFS, single macromolecules can in principle also be utilized in nanotechnology, such as for building nanoactuators, valves, or nanoscale motors.

Department of Materials Science and Technology of Polymers and MESA$^+$ Research, Institute for Nanotechnology, University of Twente, PO Box 217, 7500AE Enschede, The Netherlands.
*Email: g.j.vancso@utwente.nl

4.1. Introduction

Due to the swift development of enabling platforms, such as atomic force microscopy (AFM), in nanotechnology, visualization and spatial manipulation of single macromolecules, as well as measurements of their chain-stretching behavior, have become possible in the last few decades. Such studies are particularly important when spatial (and temporal) ensemble averaging should be avoided, molecular forces need to be measured, and individual species should be tracked. Due to the developments in AFM it has become possible to measure forces with a resolution of a few pN exerted on polymers, which were preselected by visualizing and "zooming in" on the coiled chains, as a function of chain extension. Such force–extension studies, which are of fundamental importance in macromolecular nanoscience, have opened new perspectives in life sciences, as well as in materials science.[1–3] Single chain processes that could not otherwise be directly detected, like protein folding,[4–6] DNA mechanics,[7–11] mechanical work generated by motor proteins,[12–16] identification of individual molecules,[17,18] elasticity of macromolecules,[19–24] mechanochemistry at the single bond level,[25] and binding potentials of host–guest complexes, can now be tackled.[21,26,27] AFM-based single molecule tools allow one to follow, in real time and from a true molecular perspective, the movements, forces and strains developed during the course of a reaction, as well as conformational changes induced by diverse means. The knowledge obtained by single molecule experiments is above all fundamental, and provides information complementary to existing principles, as well as encompasses quantitative information regarding interactions directly at the single chain level. In addition, the contours of applications in nanotechnology, such as single chain molecular motors to propel future nanoscale devices, single molecule valves, sensing and actuation, are gradually emerging.

The present contribution reviews the use of an approach based on AFM, namely single molecule force spectroscopy (SMFS), to perform selected mechanical experiments on individual flexible synthetic polymer chains, and the evaluation of the experimental results using the theoretical models describing single chain force–extension behavior. We highlight the tracing of external-stimulus-induced changes in single macromolecules and applications of force–extension cycles of stimulus-responsive polymers to obtain molecular motors and actuators.

4.2. AFM-Based Single Molecule Force Spectroscopy (SMFS)

The methods used for handling and testing single molecules require a probe capable of generating and measuring molecular forces and nanoscale displacements, together with the ability to spatially localize the molecules to be studied. Various techniques have been employed for single molecule manipulation and mechanical characterization, depending on the force range and minimum displacements required for the system (and problem) under study (for a review see Refs. 1 and 28). AFM-SMFS applies and senses forces through the displacement of a bendable beam (cantilever). With the advantages of high spatial sensitivity and access range, versatility, and low signal to noise, as well as the possibility of locating and probing single molecules under environmentally controlled conditions (water or organic solvent, temperature, salt, electrochemical potential, etc.), this technique has been increasingly used for the detection and mechanical characterization of single molecules. Examples are melting and unzipping forces of double-stranded DNA,[7–11] protein unfolding,[5,6,29] polysaccharide force-induced conformational transitions,[23,30–40] interfacial conformations and desorption forces,[41–51] as well as elasticity of synthetic polymers.[33,52] Intermolecular and intramolecular interactions can also be directly probed by AFM-SMFS, like the interactions between macromolecules and solvents, the effect on the statistical properties of single polymer chains,[53–56] as well as the interactions with other small molecules.[39,57–59] Examples of some of these applications are summarized in Fig. 1.

Colton *et al.* are considered the first authors to describe an experimental approach using force spectroscopy to determine the interaction between complementary strands of DNA and intrachain forces associated with the elasticity of single DNA strands.[10] In another early pioneering work, Gaub *et al.* described the use of force spectroscopy to determine intermolecular forces and energies between ligands and receptors, specifically between avidin (or streptavidin) and biotin analog.[60]

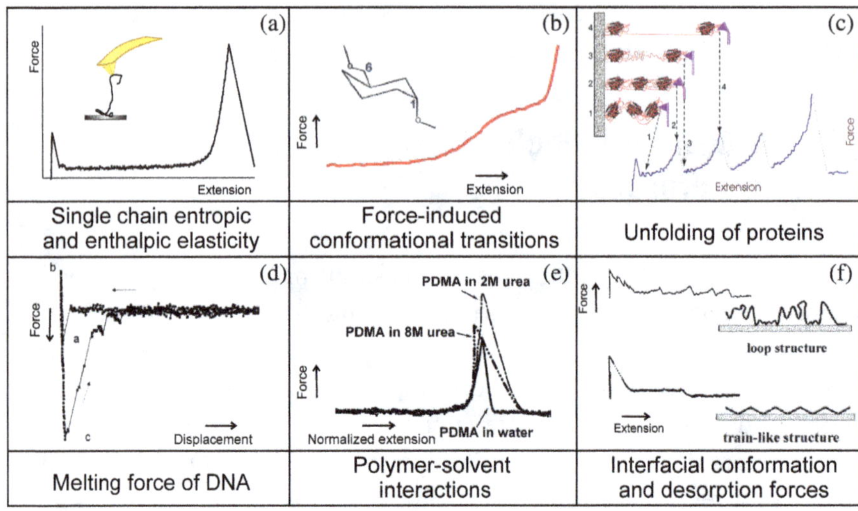

Fig. 1. Examples of the use of AFM-SMFS for studying intramolecular (a–c) and intermolecular (d–f) interactions in macromolecular systems. (a) Single chain entropic and enthalpic elasticity: schematics of a typical force–extension curve of a polymeric chain. (b) Force-induced conformational transitions: "fingerprint" of elasticity for a linear polysaccharide [dextran: α-(1,6)-D-glucopyranose] ascribed to a chair-to-boat conformational transition of the pyranose ring combined with a rotation of the exocyclic group — the simplified monomeric unit is shown[31] (*adapted by permission from Macmillan Publishers Ltd.:* Nature Biotechnology, *copyright 2001*). (c) Unfolding of proteins: the saw-tooth pattern of peaks that is observed when force is applied to extend the protein corresponds to sequential unraveling of individual domains of a modular protein[29] (*adapted from Ref. 29, with permission from Elsevier*). (d) Melting force of DNA: force versus displacement between complementary (ACTG)5 and (CAGT)5 functionalized surfaces (0.1N NaCl, pH 7.0, 25°C)[10] (*adapted from Ref. 10, with permission from AAAS*). (e) Polymer-solvent/small-molecule interactions: normalized force–extension curves of a single polydimethylacrylamide (PDMA) chain in different solvent conditions: 2 and 8M urea[57] (*reprinted in part with permission from Ref. 57, copyright 2002, American Chemical Society*). (f) Interfacial conformation and adsorption force–extension patterns for different adsorption conformations[61] (*adapted from Ref. 61, with permission from Elsevier*).

The principles of AFM-SMFS have been thoroughly described in numerous reviews.[19,21,23,26,61–65] In brief, in a typical AFM-SMFS configuration the force is sensed through the deflection of the AFM cantilever by using the optical beam principle [Fig. 2(a)]. Laser light is focused on the back of the cantilever, which is terminated by a sharp tip (with typical radius values ranging from a few to tens of nanometers) used to pick up and stretch single macromolecules. (We note that other forms of attachment, such as

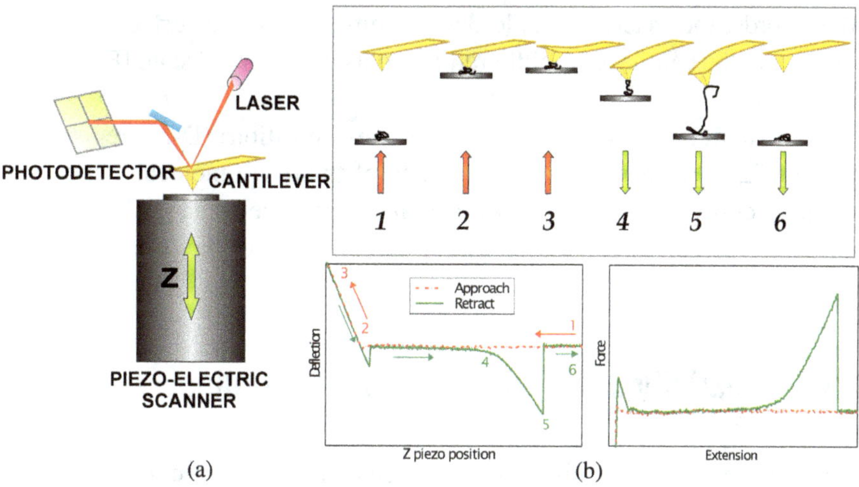

Fig. 2. (a) The principle of AFM-SMFS and (b) schematic illustration of a single molecule deflection–displacement (piezo position) experiment, the registered deflection-Z piezo position data, afterward transformed into force–extension. (*Reproduced with permission from Ref. 23.*)

chemical grafting, have been employed in addition to physisorption.) The movement of the piezoelectric scanner during an approach–retract cycle is illustrated in Fig. 2(b). During an experiment, the cantilever is initially free [Fig. 2(b), position 1]. Subsequently, the piezo, together with the substrate containing the physically or chemically adsorbed molecules, moves toward the cantilever [approach; Fig. 2(b), position 2]. (Usually the tip is stationary, and the sample is scanned.) While tip and sample are in contact, a varying force can be applied to the sample, in addition to the tip–sample interaction arising from molecular forces. The cantilever is then "pressed" to the substrate, and is thus deflected "upward" (positive deflection), and the macromolecules on the substrate can adsorb to the tip [Fig. 2(b), position 3]. Another strategy for chain attachment includes chemical tethering of the macromolecules to the AFM tip, in which case the polymer molecules can contact the substrate when the tip and the surface are in contact. Upon separation (retraction), the linking macromolecule is first uncoiled [Fig. 2(b), position 4] and stretched, which results in the deflection of the cantilever toward the substrate [Fig. 2(b), position 5]. A negative deflection is registered due to chain pulling, and as the chain is entirely stretched the weakest point of the bridging structure breaks. In other words, the macromolecule desorbs either from the surface or from the tip, and the cantilever rapidly returns to its unperturbed state [Fig. 2(b), position 6].

The deflection vs. displacement (Z piezo position) (D-z_0) registered profile [Fig. 2(b)] describes the elasticity of the macromolecule, and can be easily converted into force vs. actual distance between AFM tip and substrate (extension) (F-z) when the spring constant of the cantilever (k_s) is known.[66]

4.3. Elasticity of Individual Macromolecules

The restoring force profiles (force–extension curves) recorded during stretching experiments encompass valuable information about the elasticity of the macromolecule. A single polymer chain, which is made up of hundreds to thousands of monomers, must be treated by statistical physics due to its large number of degrees of freedom. When a single polymer chain is stretched, two kinds of restoring forces are operational. If a single chain adopts a random coil conformation, a Brownian molecular motion causes a permanent fluctuation of the molecule. Extension of the molecule leads to a reduction in the number of possible conformations, causing a loss of conformational entropy. Hence, *entropic* forces dominate at short extensions (low stretching forces). Large elongations approaching the chain contour length between the tip and the substrate, on the other hand, lead to stresses in the molecular backbone. As a result, bond angles, dihedrals and bonds become stretched and deformed, and the corresponding *enthalpic* elasticity is recorded in addition to the entropic forces.

Different statistical mechanical models can be used to describe the entropic elasticity of a polymer chain during extension. The two most frequently used ones applied in AFM-SMFS are the freely jointed chain (FJC)[67,68] and wormlike chain (WLC)[69–73] models, and their modifications (see Table 1).

In the FJC model, a polymer molecule is treated as N number of rigid Kuhn segments with lengths of l_K (Kuhn's segment length), connected through flexible joints without any long-range interactions. [Kuhn segments are useful for statistical treatments of chain molecules. A polymer can

Table 1. Elasticity models describing the force–extension relationship for single polymer chains.

Model	Force–extension expression	Fitting parameters
Freely jointed chain (FJC)[67]	$z(F) = L_c\left[\coth\left(\frac{Fl_K}{k_BT}\right) - \frac{k_BT}{Fl_K}\right]$	l_K, L_c
Wormlike chain (WLC) (Marko–Siggia)[72,73]	$F(z) = \frac{k_BT}{l_p}\left[\frac{1}{4}\left(1-\frac{z}{L_c}\right)^{-2} + \frac{z}{L_c} - \frac{1}{4}\right]$	l_p, L_c
Extended FJC (FJC$^+$ or m-FJC)[74]	$z(F) = L_c\left[\coth\left(\frac{Fl_K}{k_BT}\right) - \frac{k_BT}{Fl_K}\right]\left(1+\frac{F}{K_s l_K}\right)$	l_K, L_c, K_s
Odijk WLC[75]	$z(F) = L_c\left[1-\frac{1}{2}\left(\frac{k_BT}{Fl_p}\right)^{1/2} + \frac{F}{K}\right]$	l_p, L_c, K
Modified Marko–Siggia WLC (WLC$^+$)[76]	$F(z) = \frac{k_BT}{l_p}\left[\frac{1}{4}\left(1-\frac{z}{L_c}+\frac{F}{K_o}\right)^{-2} + \frac{z}{L_c} - \frac{F}{K_o} - \frac{1}{4}\right]$	l_p, L_c, K_o
Hooke spring modified WLC[77]	$F(z) = \frac{k_BT}{l_p}\left[\frac{1}{4}\left(1-\frac{z}{L_c}\right)^{-2} + \frac{z}{L_c} - \frac{1}{4}\right] + K'z$	l_p, L_c, K'

F: force; z: extension; k_B: Boltzmann's constant; T: temperature; L_c: contour length; l_K: Kuhn segment length; l_p: persistence length; K_s, K, K_o, K': elasticity parameters.

be divided into N physical segments (N is not necessarily the degree of polymerization), each having a length of l_K. The segments are linked in a certain way — freely jointed, freely rotating, chains of hindered rotation, etc.] The FJC model assumes that each segment has a fixed length (l_K), and the chain is fully elongated when the extension z is equal to the value Nl_K. This approach is useful for small elongations (small applied forces).[68] In order to cover the whole force (and elongation) range, a general approach was developed taking into account also the enthalpic contributions at higher elongations. This model is known as the modified or extended FJC (m-FJC or FJC$^+$)[74] and describes the molecule as n identical elastic springs in series, characterized also by the segment elasticity parameter K_s.

It is important to bear in mind that Kuhn's segment length, l_K, from the FJC model is not a fixed value for a certain molecule but includes information about the statistical configuration of a macromolecule in a specific environment. The value of l_K can be larger than the actual monomer size due to a variety of interactions among the individual monomers — for example, individual covalent bonds may have bending energies that are not small relative to k_BT. In addition, successive monomers can have steric interactions, which change the segment length, and external conditions (solvent, salt concentrations, temperature, etc.) can also alter segment length values.

On the other hand, the WLC model describes a polymer chain as a homogeneous string with a constant bending elasticity, and treats the molecule as a continuous entity with a persistence length, l_p.[69,70] This length reflects the sum of the average projections of all chain segments in a specific direction described by a given segment. If a chain is shorter than this length, the polymer is considered linear. Although this approach takes into account both entropic and enthalpic contributions, the extension is limited by the contour length of the polymer (L_c), and the model fails to describe macromolecules under conditions of high stress. (We note that for semiflexible WLC chains, the Kuhn segment length is equal to twice the value of the persistent length.) To broaden the range of applications of the WLC model, the extended WLC models by Odijk[75] (Odijk WLC) and Wang[76] (modified Marko–Siggia), among other extensions of WLC (see Table 1), have been developed.

In thermodynamic analysis of stretching, two scenarios can exist. In one case the chain ends are kept at a fixed end-to-end distance and the value of the fluctuating force is measured, for example, as a function of the end-to-end distance.

Alternatively, a fixed force can be applied while the length value of the fluctuating end-to-end distance is measured, which results in a plot of mean end-to-end distance as a function of the force. Instruments that allow one to perform the corresponding experiments are referred to as "isotensional". On the single molecule level, due to large fluctuations in the measured lengths or forces, the choice of statistical ensemble has a profound influence on the measured elastic response.[78–81] For stretching experiments using a soft cantilever (like in AFM-SMFS), the fluctuations in the entropic force and the end-to-end distance are small, and limited by the thermal fluctuations of the cantilever rather than the polymer itself. Then the thermodynamic relationship $\mathbf{F}(\mathbf{r}) = dU(\mathbf{r})/d\mathbf{r}$ is valid [for the $U(\mathbf{r})$ free energy of the chain constrained to the end-to-end distance $\mathbf{r}$]. Isotensional conditions were considered in the derivations of the force–extension relations for ideal chains (FJC and WLC).

4.3.1. *Fitting the theoretical models to the experimental data*

For a quantitative description of single chain tensile experiments, a proper theoretical model should first be chosen. In addition to selecting a proper model (or the interpolation formula for a certain model), it is crucial to make an accurate fitting of the model to the data in order to derive meaningful single chain physical parameters. Fitting is often carried out by least squares methods, and the most commonly employed one is the nonlinear, rapidly converging, regression method based on the Levenberg–Marquardt algorithm.[82]

Fitting is rather straightforward when direct force–extension functions are used, i.e. in the cases of the Marko–Siggia WLC or the Hooke spring-modified WLC interpolation formulas (see Table 1). This is due to the experimental data normally being in the form of force (F) as a function of vertical extension (z) (derived from the measured deflection versus adjusted piezo travel corrected for cantilever deflection). The fitting parameters (l_p and L_c for WLC, and also K' for the modified WLC) can thus be directly obtained. The FJC, extended FJC and Odijk WLC models, on the other hand, describe the extension as a function of force $z(F)$, i.e. the models are formulated in terms of the inverse function of the experimentally determined relationship. Consequently, before applying the fitting algorithm, an inversion of the data must be carried out. This should be accompanied by a proper statistical analysis and transformation of data statistics. Even for cases when fitting is successful,[23] attention should be paid to the validity of the mathematical method applied. The statistical noise of the experimental data for the independent variable (the force in this case) is thus transformed into the dependent coordinate. The distribution of the data points is not constant along the force regime, but is higher at low forces (mainly the entropic region), and lower in the high-force part of the curve, where the force increases much more rapidly with small extension steps. This may result in failure of the Levenberg–Marquardt fits, and appropriate weighting factors according to the error values may be considered. Occasionally, if nonphysical or poor fits are obtained, one of the parameters could be estimated and fixed (using physical clues) in order to reduce the number of fitting parameters.

Care must be taken when considering the values and physical significance of the elastic parameters obtained. As mentioned, it should be considered that the statistical models used assume a Gaussian statistical behavior of the macromolecular chains, i.e. at high extensions relative to the chain contour length between tip and substrate the models fail. Similar problems arise when the overall contour length of the polymer chain is short. Under such conditions the chain statistics cannot be assumed to be Gaussian. Moreover, the larger the number of segments (N), the softer the macromolecular "spring". This correspondence is merely a reflection of the fact that, for a given elongation, each chain segment becomes less oriented for an increasing N. In other words, short chains become highly oriented even at low strains and may cause an "overproportionate" contribution to the tensile stress.[68]

A list of the fitting parameters reported in the literature for the use of either WLC or FJC models to describe the elastic behavior of synthetic polymers is given in Ref. 23. Values of persistence length l_p are in general in the range of 0.2–0.6 nm, within the expected magnitudes for segment lengths of synthetic polymers. In some cases both models, WLC and FJC, fit the data, and the numerical values for l_p and l_K, respectively, are similar within the standard deviation, as is the case of polystyrene, poly-2-vinylpyridine and their block copolymers reported by Bemis *et al.*[83] However, a comparison of the values for different polymers, or data obtained under varying solvent conditions measured by different groups (different instruments), has a somewhat limited value. In spite of this problem, valuable "semi-quantitative" information can be obtained when the effects of solvent,[84–88] temperature[33] or other variables[43,83,87,88] are considered, especially when data are obtained from the same group and/or by using the same instruments.

4.4. Single Chain AFM Force Spectroscopy of Stimulus-Responsive Polymers

Stimuli-responsive polymers are defined as polymers that undergo relatively large and abrupt chemical or physical changes in response to small changes in the environmental conditions.[89] They recognize a stimulus as a signal and subsequently alter their chain conformation in a direct response. Synthetic polymers offer a wealth of opportunities for designing responsive materials.[90] Varying the chemical composition, length and architecture of the chains allows the response mechanisms and rates to be easily manipulated. Polymer chains may have different degrees of expansion in solution, depending on segment–segment, segment–solvent interactions, solvent partition and clustering. If size changes in the random coils are triggered by external stimuli, then the differences in free energy could be harvested as mechanical work.

In nature, several classes of molecular motors fulfill different functions in living cells, such as proton pumps in membranes,[91,92] motor proteins like myosin,[93] DNA and RNA polymerases,[94] and flagellar motors in bacteria,[95] which convert chemical energy into mechanical work. A molecular motor is a molecule that is operated in a controlled cyclic fashion to perform mechanical movement (work output) as a consequence of appropriate external stimulation (energy input).

AFM-SMFS has become a very valuable tool for characterizing conformational changes in polymer chains. The combination of this tool and the great application potential of stimuli-responsive polymers has opened up a new area of research on artificial molecular motors at the single molecule level, aimed at obtaining a fundamental understanding of the relevant molecular scale processes and at realizing and exploiting the smallest man-made machinery.[96,97]

4.4.1. *Single chain behavior of stimulus-responsive polymers*

Experimental conditions such as temperature, choice of medium and other parameters of the probing environment are usually kept unchanged in a typical AFM-SMFS test. However, for functional materials it is of interest to explore the dynamic behavior of their molecular constituents under varying environmental conditions. For example, stimulus-responsive materials hold the promise of becoming essential parts of "intelligent" devices and systems. If progress is to be made in this area, then the stimulus-responsive behavior of the constituents must be understood and eventually controlled. This need has triggered recent interest in AFM-SMFS of "smart" macromolecules, including studies of optically responsive systems[98,99] and organometallic polymers with electrochemical responsive behavior,[100,101] as well as pH- and temperature-sensitive chains. Despite this early research, the field of AFM-SMFS of such molecules is still in its infancy.

One of the most-studied stimulus-responsive synthetic polymers to date is poly(*N*-isopropyl acrylamide) (PNIPAM) [Fig. 3(a)]. PNIPAM exhibits a lower critical solution temperature (LCST) in aqueous media around 32°C.[102] As this temperature is close to the body temperature, PNIPAM has attracted interest in biomedical applications.[103] At temperatures $T < T_{\mathrm{LCST}}$ the polymer is soluble in water, owing to the dominating, solvating water–polymer segment interactions. For $T > T_{\mathrm{LCST}}$ the polymer becomes insoluble. Essential for the phase behavior of PNIPAM is the balance among the various interaction energy terms (segment–segment and segment–solvent), including specific (H-bonding) and nonspecific (van der Waals) contributions, in an aqueous environment. Increasing the temperature from below the LCST, this balance is significantly perturbed at the transition point, and a coil-to-globule transformation takes place.[104] At higher temperatures the entropy term dominates the otherwise exothermic enthalpy of the H-bonds formed between the polar groups of

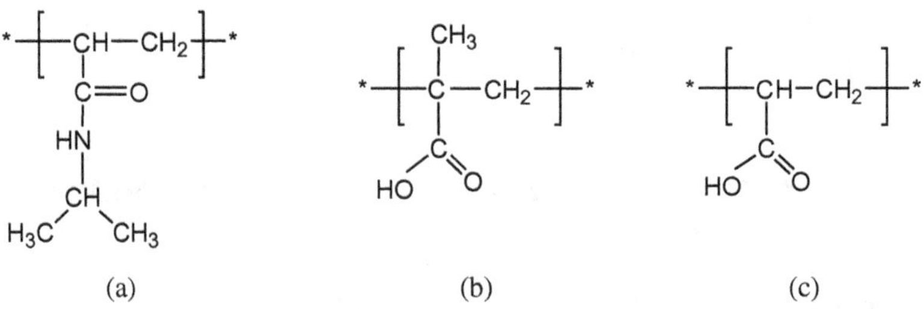

Fig. 3. The chemical structures of the stimulus-responsive polymers: (a) PNIPAM, (b) PMAA, (c) PAA.

the polymer and water molecules, which is the initial driving force for dissolution.[84] According to Tanaka *et al.*,[104] conformational rearrangements due to the enhanced segment–segment H-bonded interaction govern the formation of hydrophobic "globules," separated by solvated sections of the chain. Hence, it is of interest to observe the single chain behavior of PNIPAM "from a true molecular perspective" via single chain force spectroscopy, specifically as a function of temperature. The first single chain room temperature PNIPAM stretching experiments have been reported by Zhang *et al.*[84] The experiments were done in water and in 8 M urea solution. The force–extension curves were fitted by the m-FJC model. The value of the segment elasticity (K_s) as well as the Kuhn segment length (l_K) value increased when the solvent was changed from deionized water to 8 M urea solution, from 25 ± 2 N/m to 40 ± 5 N/m and 0.70 ± 0.05 nm to 0.78 ± 0.06 nm, respectively. Additionally, the authors heated the samples to 33°C, kept them for 10 min and then performed the original RT experiments — but, unfortunately, numerical fitting, regarding values of chain-statistical parameters, was inconclusive. According to their explanation, during the treatment the adsorption of the chains to the substrate changed. This was considered as a severe limiting factor to reproduce initial force curves (prior to temperature treatment). This issue is indicative of the great importance of chain attachment conformation, segment arrangement, and physisorption to the tip during such a "fishing" type of AFM-SMFS experiments, i.e. when no tethering of the chain to the AFM tip is done. The question thus remains open whether the physisorbed chains statistically regained their conformation following the heating treatment. In a short communication elsewhere we have described T-dependent single chain stretching experiments using AFM-tip-grafted chains.[105]

In short, we performed variable temperature single chain AFM force spectroscopy experiments on PNIPAM to elucidate the role of H-bonding in the stiffness of isolated macromolecules. In these experiments, PNIPAM chains were grafted to Au-coated AFM tips using thiol-functionalized chains, and their stretching behavior was monitored. No change in the single chain force–extension curves below and above the LCST was found. This result indicates that for PNIPAM chains under mechanical stress the formation of intrachain H-bonds at $T > T_{LCST}$ was suppressed.

Analogous medium change-driven responsive behavior can be triggered by e.g. variations of concentration, solvent quality or solvent pH (for polyionic chains). For instance, poly(acrylic acid) (PAA) and poly(methacrylic acid) (PMAA) [Fig. 3(b)] become charged by deprotonation at pH > pK_a. Such pH-responsive systems have drawn attention as chemical actuators in responsive hydrogels.[106,90]

Polymer brushes obtained using "grafting from" approaches in combination with controlled polymerizations have recently attracted great interest, specifically if they consist of stimulus-responsive chains.[107,108] A notoriously difficult analytical question in characterizing polymer brushes is related to the determination of the chain length values in the brush.[109] In a recent work by Ryan *et al.*[110] PMAA brushes and chain length of macromolecules obtained using a RAFT polymerization and "grafting from" approach were characterized by AFM-"SMFS." (The quotation marks here refer to the still-open question whether the corresponding AFM chain-stretching experiments were indeed performed on single chains.) The measured single molecule force–extension curves were fitted using the WLC model. The values of the contour length (L_c) of the polymer were easily obtained from the fitting, as the authors used a fixed persistence length (l_p) value (of 0.5 nm, in deionized water) adapted from the literature, which was measured by small-angle X-ray scattering. In principle, the value of l_p can be obtained from force curve fits. However, due to (a) experimental noise, (b) uncertainties of the "zero" point of the stretching curve, (c) the level of zero force and (d) uncertainties of the value of the extended chain length of the given macromolecule stretched, the numerical values derived by fitting all relevant parameters in the various single chain models in AFM-SMFS force–extension curves may be plagued by significant error. By assuming a physically substantiated value for l_p, the authors circumvented in part these difficulties. In contrast, in studies by Hadziioannou *et al.*,[111] using individually grafted chains, and performing experiments under the same conditions as Ryan *et al.*,[110] obtained a value of the persistence length by a fitting of 0.28 ± 0.05 nm using the WLC model. We note that both groups reported an accurate fitting.

As mentioned, by determining the distribution of chain contour lengths using AFM-SMFS on polymer brushes, one can generally obtain the chain length (and thus the molar mass) distribution in the brush.[110] To calculate the molar mass of grafted chains, Ryan *et al.* used a simple equation,

$$M_i = \frac{M_0 L_{c,i}}{l_p} \tag{1}$$

where M_i is the molar mass of the "ith" chain, $L_{c,i}$ is its contour length and M_0 is the molar mass of the monomer units. Thus, the error in determining l_p is carried over to M_i.

When we stretch a polymer chain in repeated "fishing" experiments (chains attached to a substrate), the section of the chain being stretched between tip and substrate is actually a statistical variable as the adhesive contact between tip and chain is established at random segment locations. This adds uncertainty to the contour length values obtained.

Hadziioannou *et al.* used thiol-terminated PMAA (HS-PMAA) chains grafted onto a gold substrate surface[111] with a low-grafting-density surface in order to increase the chance of single chain-stretching events. The experimental data were fitted by both the FJC and the WLC model. The l_K values derived had a magnitude of 0.33 ± 0.05 nm, in the same range as the values of l_p, equal to 0.28 ± 0.05 nm. According to the FJC and WLC models, this virtual agreement raises a fundamental question. Under the experimental conditions used, and assuming that the models are valid, the value of l_p should be half that of l_K.[112] We note that, nevertheless, the L_c value obtained in this study is in good agreement with the chain length derived from the gel permeation chromatography experiment.

Regarding the chain chemical structure, PAA [Fig. 3(c)] differs from PMAA only by one methyl group substituent. Due to the similarity in the chemical structures, one would expect similar stretching properties. In a study reported by Li *et al.*, PAA force–extension curves were fitted using the FJC and the m-FJC model. It was shown that the macromolecular stretching is not just entropy-controlled, as the use of the simple FJC model did not provide an accurate fitting.[113] The resulting value for the Kuhn segment length was equal to 0.64 ± 0.05 nm for both FJC and m-FJC.

As mentioned earlier, the properties of the solvent (hydrophilicity, ionic strength, pH) have a great influence on the l_p and the l_K values, as summarized in Ref. 23, for example. This may account for the observed different behavior of the two polymers as PAA stretching experiments were performed in 10^{-3} M KNO_3 solution, whereas the PMAA testing was conducted in deionized water. Another noteworthy point is related to the fitting quality. As mentioned, while for PMAA the simple FJC model described well the stretching, the fitting quality was poor for PAA. This difference was assigned to enthalpic contributions, which must be considered for higher stretching forces.[113] The choice of a threshold force value above which enthalpic contributions must be considered remains somewhat arbitrary, depending also on the primary structure of the macromolecule. However, as a reasonable approximation, one can assume that tensions at forces below *ca.* 20 pN do not cause torsions in bonding angles that would necessitate the use of enthalpic terms.[26] We note that for an elastic rod model, Odijk put the threshold value at 10 pN,[75] although he mentioned that the regime that demarcates the entropy from enthalpy dominance is not sharply defined (no pure entropic or enthalpic domains exist).

Reversible conformational changes and stimulus-responsive behavior can also be triggered by light.[98,99] In particular, Gaub *et al.*, in a series of articles, described detailed experiments on a polyazopeptide depicted in Fig. 4(a). Azobenzenes are well-known chromophores, which can exhibit either *trans* or *cis* conformations that can be reversibly switched with light of the appropriate wavelength. The energy barrier separating the two conformers has a height of 40 $k_B T$, with the *trans* state having a lower free enthalpy. *Cis–trans* isomerization can be achieved by optical excitation at a wavelength of 420 nm via the first excited singlet state. A *trans–cis* transition via the same excited state can in turn be triggered at a somewhat higher energy corresponding to 365 nm wavelength radiation. AFM-SMFS experiments

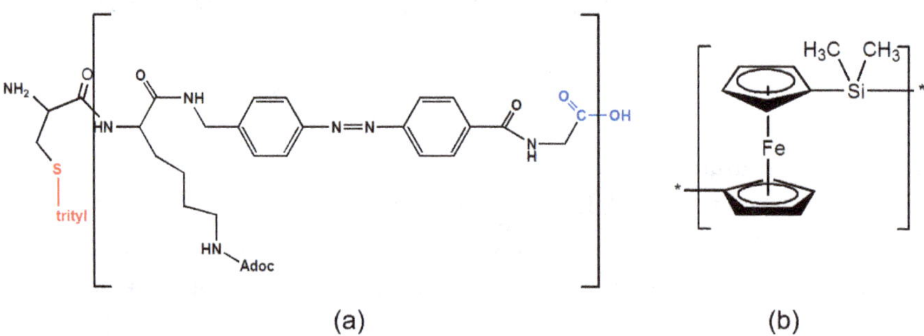

Fig. 4. The chemical structure of the azobenzene-containing polypeptide chain[98] (a) and dimethyl-substituted PFS (which can be end-capped by ethylene sulfide[96]) (b).

demonstrated that chains with *cis* isomers in their backbone remain stable for all experimentally accessible pulling forces, within the experimentally accessible timescale (the timescale is relevant due to the loading rate dependence of transition forces).[114] Gaub *et al.* exploited the bistability and reversible transitions of the polyazopeptide for constructing a "macromolecular motor" which converted optical energy into mechanical work in "optomechanical" cycles. The working principles of this molecular motor will be elucidated later.

Initially, single chain behavior of the polyazopeptide was fitted by the modified Marko–Siggia WLC model (see Table 1). The fitting parameters included l_p, K_o and L_c. The values of l_p and K_o were first determined with "reasonable" WLC fits. The contour lengths of the chains stretched in a given experiment were then calculated using these values. The authors assumed that the *cis–trans* isomerization did not change the persistence length and the segment elasticity. In a subsequent study they extended their analysis to include a freely rotating chain model, combined with quantum-mechanical calculations (FRCQM):

$$z(F) = L_c(N_p)\left[1 - \left(\frac{2Fb_p}{k_bT}\right)^{-1}\right] \tag{2}$$

Here b_p corresponds to the length of a chemical bond, N_p is the degree of polymerization and L_c is the contour length. In the azobenzene peptide chain there are three structural units, corresponding to three building blocks, i.e. the *cis*-azobenzene, the *trans*-azobenzene and the tripeptide (see structure in Fig. 4). Under tension the effective bond lengths vary, and this variation for the three types of bonds was calculated by *abinitio* methods as a function of force (QM modification).[115] Using this method the number of *cis* and *trans* units along the chain could be estimated. Such estimation was employed to assess the optomechanical conversion efficiency of isomerization processes when the chain was irradiated to induce *cis–trans* (or *trans–cis*) isomerization. Results obtained also explained why the thermal relaxation of the *cis* conformer did not speed up under tension, thus explaining why forces up to 500 pN did not change the partition of the *cis* and *trans* conformations during experiments with lengths of several seconds.

The last group of stimulus-responsive polymers that are discussed here belong to poly(ferrocenyl silanes) (PFSs) (Fig. 4(b)). These polymers belong to the group of organometallic macromolecules featuring a backbone consisting of ferrocenyl units connected by substituted silanes. Symmetric and asymmetric substitutions have been described, as well as ionic substituent groups which render PFSs water-soluble.[116] The choice of the substituents, i.e. the primary chemical structure of the chain, controls the physical properties of the polymer. Due to the presence of ferrocene in the main chain, PFSs are electrochemically active, and can be reversibly oxidized/reduced. It should be mentioned that most other redox-responsive macromolecules encompass electroactive (conducting) polymers, which are often intractable and oxygen-sensitive, which limits their applicability. During oxidation the main chain of PFS becomes positively charged, which results in a change in the stiffness. Originally hydrophobic (neutral) PFS derivatives become more polar as a result of oxidation and the oxidized polymer becomes poorly soluble in organic solvents. Further unique, and highly interesting, properties of PFS concerning applications include high etch resistivity in reactive ion etching environments (with relevance to lithography), catalytic activity, sensing, electrochromic materials, etc. Prior to exploring these applications, knowledge of single chain properties (especially for nanotechnology applications) is very valuable.

Zhang *et al.*[100] studied the elastic properties of two different PFS derivatives, poly(ferrocenyl dimethylsilane) (PFS) and poly(ferrocenyl methylphenylsilane) (PFMPS), as well as their chemically (with $FeCl_3$) oxidized forms. SMFS force–extension curves were measured in a droplet of tetrahydrofuran (THF). Fitting by the m-FJC model resulted in different values of l_K and K_s, depending on the degree of oxidation and substitution. By changing one of the methyl groups to a phenyl group, the l_K fitted value decreased (from 0.41 ± 0.01 nm to 0.36 ± 0.01 nm) but the K_s value increased from 53 ± 1 N/m (PFS) to 61 ± 1 N/m (PFMPS). Oxidizing the polymer made a more significant change in K_s (115 ± 1 N/m for PFS; 500 ± 1 N/m for PFMPS), while the increase in the value of l_k was relatively less significant (0.46 ± 0.01 nm for PFS; 0.40 ± 0.01 nm for PFMPS).

Vancso *et al.* described AFM-SMFS on PFS in a subsequent study under electrochemical control.[88,101] In their experiments ethylene sulfide end-capped poly(ferrocenyl dimethylsilane) $(PFS)_{100}$ (the subscript refers to the number average degree of polymerization) was covalently attached by grafting onto a gold surface, through inserting the chain into the defects of a self-assembled, inert monolayer. This allowed studies of separated, single chains. In order to compare differences of elasticity between neutral and oxidized PFS macromolecules under electrochemical control, individual chains were probed by AFM-SMFS. Since the targeted electrochemistry experiments were carried out in aqueous solutions, the single chain stretching of neutral PFS was also performed in aqueous solutions of $NaClO_4$ for comparison. From the fits of the force–extension curves, it can be concluded that the m-FJC model describes both entropic (both the low- and the high-force regime) and enthalpic (high-force regime) elasticities of all types of PFS polymer chains very well. In this work, an increase in l_k (from 0.40 nm in the neutral state to 0.65 nm in the oxidized state) of the PFS was observed, which was attributed to electrostatic repulsion. This contribution originates from the positive charges that are distributed along the polymer chains. According to the classical Odijk–Skolnick–Fixman (OSF) theory,[117,118] it is expected that the electrostatic interactions between the charges along the polymer chain increase the distances between like-charged segments, i.e. the stretched conformation of the chains is favored. The resulting increase in the Kuhn segment length corresponds to a lower restoring force or elasticity of the oxidized PFS. We note that incomplete oxidation by chemical oxidants caused a smaller variation of the single chain statistical parameters, as compared to full electrochemical oxidation.

Zhang *et al.*,[100] who used $FeCl_3$ as chemical oxidation agent for PFS homopolymers, reported a substantial increase in the enthalpic elasticity for PFS. The Kuhn segment length increased by 10% following oxidation, and the value of the segment elasticity was more than doubled. These data indicate, in agreement with the data of Vancso's group, that the elasticity of individual PFS macromolecules was changed after oxidation.

In addition to fundamental scientific interest in single chain behavior, there are potentially relevant technological consequences of AFM-SMFS for macromolecules. For example, for future functional nanoscale devices, such as pumps, valves, levers and molecular walkers, molecular scale motors will be required to power these structures. Moreover, a complete understanding of the working principles of biological molecular motors, e.g. of motor proteins, and their use in artificial muscles, need further research. These issues have motivated work aimed at single chain molecular motors, which will be discussed in the next section.

4.4.2. *Single molecule optomechanical cycle*

A first demonstration of photomechanical energy conversion in an individual molecule was reported by the group of Gaub,[98,99] who showed reversible, optical switching of individual molecules of a polymer containing azobenzene groups in the backbone, i.e. a polyazopeptide (see Fig. 4). The group demonstrated that the contour length of the polymer was selectively lengthened or shortened by switching between *trans*- and *cis*-azo configurations, when applying wavelengths of specific ultraviolet light (see below). Changes in contour length were reported, both at low forces and under external loads of up to 400 pN. It was demonstrated that the mechanical stability of the two azobenzene configurations is sufficient to operate the experiment in an optomechanical cycle (as will be discussed later) and thus to perform work at the molecular level.

Figure 5 shows a scheme of the experimental setup used, and the change in the force–extension response of the optically excited and nonexcited molecules. The experiment started with a polymer in a nondefined, initial, configurationally mixed state (black trace; $L = 86.5$ nm). By the application of five 420 nm pulses, the polymer was switched to the saturated *trans* state and lengthened by *ca.* 1.4 nm (red trace). After five pulses at 365 nm, the same molecule was shortened by $\Delta L = 2.8$ nm. In order to obtain the contour lengths L of the stretched polymer molecules, the force–extension data were fitted with the extended WLC model. These results are shown in Fig. 5 as the traces to the right. In the traces to the left, a single polyazopeptide strand was shortened against an external force. The strand was first driven into the *trans* state by five pulses at 420 nm, after which it was stretched to a force of about 350 pN (red trace). The application of one pulse at 365 nm at a constant tip/sample separation resulted in a shortening of *ca.* 1.1 nm (black trace). Two further pulses resulted in an additional shortening by 0.8 nm (blue trace). Neither of five additional pulses was found to result in any further shortening, and thus the polymer was assumed to be in the saturated *cis* state.

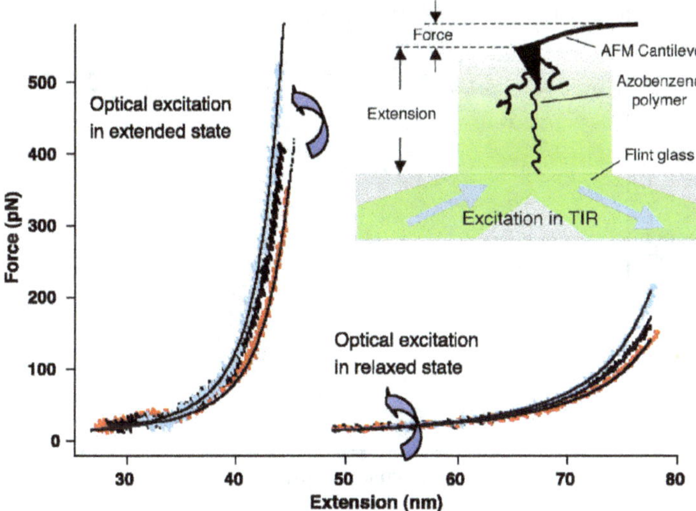

Fig. 5. The traces to the right portray the force–extension of a single polyazopeptide. The traces to the left correspond to a single polyazopeptide strand being shortened against an external force. Inset: Schematic of the experimental setup. TIR: total internal reflection. (*Reprinted from Ref. 99, with permission from AAAS.*)

The closing of a cycle was carried out in the way portrayed in Fig. 6: an individual azopolymer was first optically lengthened (pulse application) (1), and then mechanically expanded to a certain restoring force (2). The application of a new pulse contracted the polymer against the external force (3), and finally the force was again reduced (4). The cycle was completed by the optical expansion of the molecule to its original state. Since the mechanical work at the molecular level was a result of a macroscopic optical excitation, the real quantum efficiency of optomechanical switching for the cycle in the AFM setup was only on the order of 10^{-18}. However, the maximum efficiency of the optomechanical energy conversion at a molecular level was estimated to be about 10%.

In the extended state, the shortening transition was able to convert the absorbed optical energy into mechanical work performed on the system. This was the first demonstration that such light-sensitive molecules could be used for a molecular switch via length change or to perform work on (or by) a single chain system.

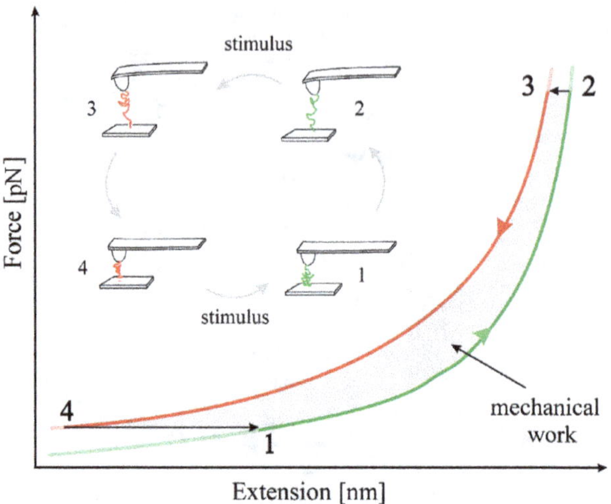

Fig. 6. A schematic of a single molecule operating cycle.

4.4.3. *Realization of a redox-driven single macromolecule motor*

Recently, Vancso and coworkers have systematically investigated the stimulus-responsive poly(ferrocenyl dimethylsilane) (PFS) polymer[119,120] as a model system for the realization of (macro)molecular motors powered by a redox process. This is an alternative way to switch the flexibility of the chain, where addressing single chains can in principle be confined to even a single macromolecule.[96,99,121] AFM-SMFS experiments described in the previous section provided the basis for the demonstration of a single macromolecule motor powered by redox energy.

A proposed possible experimental cycle to realize the molecular motor was first defined by keeping the deflection of the cantilever constant in AFM-SMFS measurements, i.e. constant force, during the transition from the oxidized to the neutral state (and vice versa) (Fig. 6). The two branches (neutral or oxidized) were determined by the corresponding elasticities of the polymer, well described by the m-FJC model. Starting from a low force (20 pN) (1 in Fig. 6) under an applied constant external potential of +0.5 V, an individual, oxidized PFS polymer chain with 50 nm contour length is pulled to a force of 140 pN (2). At a constant force of 140 pN, the PFS chain is reduced to its neutral state by controlling the external potential back to 0 V (3), thus giving rise to a change in the elasticity of the polymer chain. Subsequently, the force on the polymer is reduced back to 20 pN (4) and, finally, the cycle is completed by applying an external potential of +0.5 V to completely oxidize the whole PFS chain. By periodically controlling the external potential, the corresponding oxidized and/or neutral PFS chains can be created to realize the operating cycle, with a mechanical work of about 3.4×10^{-19} J, as calculated from the enclosed area.

Such electrochemically driven, macromolecule-based motors are potentially interesting for the realization of single molecule devices, as they can in principle be addressed on the single molecule level by using miniaturized electrodes (for PFS) and can be repeatedly run in cycles in a reversible manner. In this context, it is important to determine how the efficiency of single motors depends on various experimental and (macro)molecular design parameters. This necessitates the analysis of closed mechanoelectrochemical cycles of individual macromolecules, including the localization and addressing of a single macromolecule by the AFM tip, and the stretching and relaxing of the molecule *in situ* under different applied electrochemical potentials.

The first experimental realization of closed mechanoelectrochemical cycles of individual PFS chains in electrochemical AFM-based SMFS was reported by Vancso and coworkers.[96] Individual PFS chains kept in an extended state between the AFM tip and the electrode surface were electrochemically oxidized at a constant z position by applying a potential of +0.5 V [cycle 1, 1-2-3-1′, as shown in Fig. 7(a)]. Cyclic voltammetry was performed to ensure complete oxidation of the PFS chains on the gold working electrode. Correspondingly, single chains were stretched in the oxidized state (by keeping the potential at +0.5 V), followed by electrochemical reduction at a constant z position and continued force spectroscopy [cycle 2, 1-3-2-1′, as shown in Fig. 7(b)]. The two data sets display the entire mechanoelectrochemical experimental cycles for the individual molecules in both possible directions.

In the experiments, the force at fixed maximum extension was observed to decrease upon oxidation and increase upon reduction to the neutral state due to the lengthening of the oxidized chain with respect to the neutral one. This is attributed to the electrostatic repulsion among the

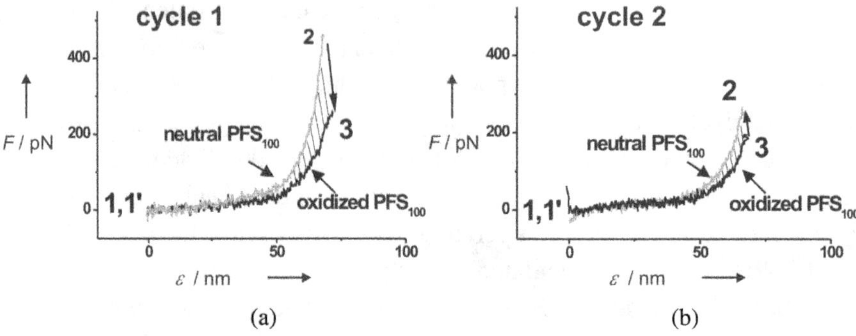

Fig. 7. Force–extension curves of (a) cycle 1 and (b) cycle 2. The enclosed areas of the cycles correspond to the mechanical work input or output of the single polymer chain mechanoelectrochemical cycle. (*Adapted from Ref. 96, with permission from Wiley-VCH.*)

oxidized ferrocene centers along the chain.[75,117] Therefore, the change in the redox state is directly coupled to a mechanical output signal of the force sensor (for cycle 1: ≈200 pN).

4.5. Conclusions and Outlook

This chapter has demonstrated the use of AFM-based single molecule force spectroscopy to perform molecular tensile experiments on individual flexible synthetic polymer chains in different environments, and the evaluation of the experimental results using the theoretical models for stretching of macromolecules. In the discussions stimulus-responsive polymers and their potential as molecular nanomotors received particular attention. Variations in chain segment and elasticity parameters have been derived quantitatively from single chain statistical models. Due to open problems in force–chain extension data fitting, the "ill-posed" mathematical problem of the fitting itself (related to experimental noise), as well as uncertainties in obtaining contour lengths and zero force levels (experimentally), mechanical creep and other instrumental problems, the numerical results can at present be considered only as semiquantitative.

The interpretation of AFM-SMFS data by single polymer chain models also needs attention. Classical models, like the WLC and FJC models, describe stretching experiments reasonably well. These models have been extended to fit the force–extension curves also in the high-force (enthalpic) regime. Meanwhile, quantum-mechanical approaches, like density function theory, *ab initio* calculations, as well as molecular dynamics and Monte Carlo simulations, have been introduced. The development of more advanced data treatment procedures, standardization of the data processing and analysis, and proper statistical evaluations, in combination with the new theories, are needed for further progress.

The ability of AFM-SMFS to sense conformational changes in single macromolecules, and the feasibility of the conversion of the externally applied energy into mechanical work, have been demonstrated in the review. These results represent a significant step in the development of nanoactuators and synthetic molecular motors. It has also been shown that the maximum attainable efficiency of a motor cycle can be enhanced by proper molecular design of the macromolecule. Localized excitation, improved thermal stability and elimination of all sources of cantilever drift are experimental parameters that need further improvement. However, progress has clearly demonstrated that in addition to fundamental scientific interest in single chain behavior, there are potentially relevant technological consequences of AFM-SMFS for macromolecules.

Acknowledgments

The Netherlands Organization for Scientific Research (NWO TOP Grant 700.56.322, Macromolecular Nanotechnology with Stimulus-Responsive Polymers; and NWO Equipment Grant) is gratefully acknowledged for financial support.

References

1. C. Bustamante, J. C. Macosko and G. J. L. Wuite, *Nat. Rev. Mol. Cell Biol.* **1**, 130 (2000).
2. A. L. Weisenhorn, P. K. Hansma, T. R. Albrecht and C. F. Quate, *Appl. Phys. Lett.* **54**, 2651 (1989).

3. P. Williams, in *Scanning Probe Microscopies Beyond Imaging: Manipulation of Molecules and Nanostructures*, ed. P. Samorì (Wiley-VCH, Weinheim, 2006), pp. 250–274.
4. M. Carrion-Vazquez, A. F. Oberhauser, S. B. Fowler, P. E. Marszalek, S. E. Broedel, J. Clarke and J. M. Fernandez, *Proc. Natl. Acad. Sci. USA*, **96**, 3694 (1999).
5. H. Li, W. A. Linke, A. F. Oberhauser, M. Carrion-Vazquez, J. G. Kerkvliet, H. Lu, P. E. Marszalek and J. M. Fernandez, *Nature* **418**, 998 (2002).
6. K. Mitsui, M. Hara and A. Ikai, *FEBS Lett.* **385**, 29 (1996).
7. C. Bustamante, Z. Bryant and S. B. Smith, *Nature* **421**, 423 (2003).
8. R. Krautbauer, H. Clausen-Schaumann and H. E. Gaub, *Angew. Chem. Int. Ed.* **39**, 3912 (2000).
9. R. Krautbauer, M. Rief and H. E. Gaub, *Nano Lett.* **3**, 493 (2003).
10. G. U. Lee, L. A. Chrisey and R. J. Colton, *Science* **266**, 771 (1994).
11. M. Rief, H. Clausen-Schaumann and H. E. Gaub, *Nat. Struct. Biol.* **6**, 346 (1999).
12. C. Bustamante, D. Keller and G.Oster, *Acc. Chem. Res.* **34**, 412 (2001).
13. D. Keller and C. Bustamante, *Biophys. J.* **78**, 541 (2000).
14. A. Kishino and T. Yanagida, *Nature* **334**, 74 (1988).
15. I. Rayment, H. M. Holden, M. Whittaker, C. B. Yohn, M. Lorenz, K. C. Holmes and R. Milligan, *Science* **261**, 58 (1993).
16. R. D. Vale and R. A. Milligan, *Science* **288**, 88 (2000).
17. M. S. Z. Kellermayer, *Physiol. Meas.* **26**, R119 (2005).
18. P. Samorì, M. Surin, V. Palermo, R. Lazzaroni and P. Leclère, *Phys. Chem. Chem. Phys.* **8**, 3927 (2006).
19. A. Janshoff, M. Neitzert, Y. Oberdörfer and H. Fuchs, *Angew. Chem. Int. Ed.* **39**, 3213 (2000).
20. T. R. Strick, M. N. Dessinges, G. Charvin, N. H. Dekker, J. F. Allemand, D. Bensimon and V. Croquette, *Rep. Prog. Phys.* **66**, 1 (2003).
21. J. Zlatanova, S. M. Lindsay and S. H. Leuba, *Prog. Biophys. Mol. Biol.* **74**, 37 (2000).
22. M. I. Giannotti, W. Shi, S. Zou, H. Schönherr and G. J. Vancso, in *Surface Design: Applications in Bioscience and Nanotechnology*, ed. R. Förch (Wiley-VCH, Weinheim, 2009), pp. 405–427.
23. M. I. Giannotti and G. J. Vancso, *Chem. Phys. Chem.* **8**, 2290 (2007).
24. X. Zhang, C. J. Liu and Z. Q. Wang, *Polymer* **49**, 3353 (2008).
25. S. Garcia-Manyes, J. Liang, R. Szoszkiewicz, T. L. Kuo and J. M. Fernandez, *Nat. Chem.* **1**, 236 (2009).
26. T. Hugel and M. Seitz, *Macromol. Rapid. Commun.* **22**, 989 (2001).
27. S. Zou, H. Schönherr and G. J.Vancso, in *Scanning Probe Microscopies Beyond Imaging: Manipulation of Molecules and Nanostructures*, ed. P. Samorì (Wiley-VCH, Weinheim, 2006), pp. 315–353.
28. H. Clausen-Schaumann, M. Seitz, R. Krautbauer and H. E. Gaub, *Curr. Opin. Chem. Biol.* **4**, 524 (2000).

29. T. E. Fisher, A. F. Oberhauser, M. Carrion-Vazquez, P. E. Marszalek and J. M. Fernandez, *Trends Biochem. Sci.* **24**, 379 (1999).
30. M. Rief, F. Oesterhelt, B. Heymann and H. E. Gaub, *Science* **275**, 1295 (1997).
31. P. E. Marszalek, H. B. Li and J. M. Fernandez, *Nat. Biotech.* **19**, 258 (2001).
32. N. I. Abu-Lail and T. A. Camesano, *J. Microscopy* **212**, 217 (2003).
33. M. I. Giannotti, M. Rinaudo and G. J. Vancso, *Biomacromolecules* **8**, 2648 (2007).
34. R. G. Haverkamp, A. T. Marshall and M. A. K. Williams, *Phys. Rev. E: Stat. Nonlin. Soft Matter Phys.* **75**, (2p+1) 021907 (2007).
35. H. B. Li, M. Rief, F. Oesterhelt, H. E. Gaub, X. Zhang and J. C. Shen, *Chem. Phys. Lett.* **305**, 197 (1999).
36. P. E. Marszalek, A. F. Oberhauser, Y. P. Pang and J. M. Fernandez, *Nature* **396**, 661 (1998).
37. K. A. Walther, J. Brujic, H. B. Li and J. M. Fernandez, *Biophys. J.* **90**, 3806 (2006).
38. Q. B. Xu, W. Zhang and X. Zhang, *Macromolecules* **35**, 871 (2002).
39. Q. M. Zhang, Z. Y. Lu, H. Hu, W. T. Yang and P. E. Marszalek, *J. Am. Chem. Soc.* **128**, 9387 (2006).
40. Q. M. Zhang and P. E. Marszalek, *J. Am. Chem. Soc.* **128**, 5596 (2006).
41. X. Chatellier, T. J. Senden, J. F. Joanny and J. M. di Meglio, *Europhys. Lett.* **41**, 303 (1998).
42. M. Conti, Y. Bustanji, G. Falini, P. Ferruti, S. Stefoni and B. Samori, *Chem. Phys. Chem.* **2**, 610 (2001).
43. S. X. Cui, C. J. Liu, Z. Q. Wang and X. Zhang, *Macromolecules* **37**, 946 (2004).
44. S. X. Cui, C. J. Liu, W. Zhang, X. Zhang and C. Wu, *Macromolecules* **36**, 3779 (2003).
45. S. X. Cui, C. J. Liu and X. Zhang, *Nano Lett.* **3**, 245 (2003).
46. C. Friedsam, A. D. Becares, U. Jonas, H. E. Gaub and M. Seitz, *Chem. Phys. Chem.* **5**, 388 (2004).
47. C. Friedsam, A. D. Becares, U. Jonas, M. Seitz and H. E. Gaub, *New J. Phys.* **6**, 9 (2004).
48. L. Garnier, B. Gauthier-Manuel, E. W. van der Vegte, J. Snijders and G. Hadziioannou, *J. Chem. Phys.* **113**, 2497 (2000).
49. T. Hugel, M. Grosholz, H. Clausen-Schaumann, A. Pfau, H. E. Gaub and M. Seitz, *Macromolecules* **34**, 1039 (2001).
50. M. Seitz, C. Friedsam, W. Jostl, T. Hugel and H. E. Gaub, *Chem. Phys. Chem.* **4**, 986 (2003).
51. L. Sonnenberg, J. Parvole, O. Borisov, L. Billon, H. E. Gaub and M. Seitz, *Macromolecules* **39**, 281 (2006).
52. C. J. Liu, W. Q. Shi, S. X. Cui, Z. Q. Wang and X. Zhang, *Curr. Opin. Solid State Mater. Sci.* **9**, 140 (2005).
53. H. B. Li, W. K. Zhang, W. Q. Xu and X. Zhang, *Macromolecules* **33**, 465 (2000).
54. H. B. Li, W. K. Zhang, X. Zhang, J. C. Shen, B. B. Liu, C. X. Gao and G. T. Zou, *Macromol. Rapid Comm.* **19**, 609 (1998).

55. C. J. Liu, S. X. Cui, Z. Q. Wang and X. Zhang, *J. Phys. Chem. B* **109**, 14807 (2005).
56. F. Oesterhelt, M. Rief and H. E. Gaub, *New J. Phys.* **1**, 6 (1999).
57. C. Wang, W. Q. Shi, W. K. Zhang, X. Zhang, Y. Katsumoto and Y. Ozaki, *Nano Lett.* **2**, 1169 (2002).
58. Q. B. Xu, S. Zou, W. K. Zhang and X. Zhang, *Macromol. Rapid Comm.* **22**, 1163 (2001).
59. S. Zou, W. K. Zhang, X. Zhang and B. Z. Jiang, *Langmuir* **17**, 4799 (2001).
60. F. Oesterhelt, D. Oesterhelt, M. Pfeiffer, A. Engel, H. E. Gaub and D. J. Müller, *Science* **288**, 143 (2000).
61. W. Zhang and X. Zhang, *Prog. Polymer Sci.* **28**, 1271 (2003).
62. R. Merkel, *Phys. Rep.* **346**, 343 (2001).
63. M. Carrion-Vazquez, A. F. Oberhauser, T. E. Fisher, P. E. Marszalek, H. B. Li and J. M. Fernandez, *Prog. Biophys. Mol. Biol.* **74**, 63 (2000).
64. K. Wang, J. G. Forbes and A. J. Jin, *Prog. Biophys. Mol. Biol.* **77**, 1 (2001).
65. H. J. Butt, B. Cappella and M. Kappl, *Surf. Sci. Rep.* **59**, 1 (2005).
66. C. S. Hodges, *Adv. Colloid Interface Sci.* **99**, 13 (2002).
67. S. B. Smith, L. Finzi and C. Bustamante, *Science* **258**, 1122 (1992).
68. F. Bueche, *Physical Properties of Polymers* (Interscience, New York, 1962).
69. G. Porod, *Monatsh. Chem.* **80**, 251 (1949).
70. O. Kratky and G. Porod, *Recl. Trav. Chim.* **68**, 1106 (1949).
71. P. J. Flory, *Statistical Mechanics of Chain Molecules* (Hanser-Gardner Publications, Munich, 1989).
72. C. Bustamante, J. F. Marko, E. D. Siggia and S. Smith, *Science* **265**, 1599 (1994).
73. J. F. Marko and E. D. Siggia, *Macromolecules* **28**, 8759 (1995).
74. S. B. Smith, Y. Cui and C. Bustamante, *Science* **271**, 795 (1996).
75. T. Odijk, *Macromolecules* **28**, 7016 (1995).
76. M. D. Wang, H. Yin, R. Landick, J. Gelles and S. M. Block, *Biophys. J.* **72**, 1335 (1997).
77. B. Zhang and J. S. Evans, *Biophys. J.* **80**, 597 (2001).
78. H. J. Kreuzer, S. H. Payne and L. Livadaru, *Biophys. J.* **80**, 2505 (2001).
79. D. E. Makarov, Z. Wang, J. B. Thompson and H. G. Hansma, *J. Chem. Phys.* **116**, 7760 (2002).
80. R. G. Winkler, *J. Chem. Phys.* **118**, 2919 (2003).
81. P. Cifra and T. Bleha, *Polymer* **48**, 2444 (2007).
82. H. W. Press, B. P. Flannery, S. A. Teukolsky and W. T. Vetterling, in *Numerical Recipes: The Art of Scientific Computing* (Cambridge University Press, 1986), p. 523ff.
83. J. E. Bemis, B. B. Akhremitchev and G. C. Walker, *Langmuir* **15**, 2799 (1999).
84. W. Zhang, S. Zou, C. Wang and X. Zhang, *J. Phys. Chem. B* **104**, 10258 (2000).
85. D. Zhang and C. Ortiz, *Macromolecules* **37**, 4271 (2004).
86. A. J. Ryan, C. J. Crook, J. R. Howse, P. Topham, R. A. L. Jones, M. Geoghegan, A. J. Parnell, L. Ruiz-Pérez, S. J. Martin, A. Cadby, J. R. P. Webster, A. J. Gleeson and W. Bras, *Faraday Discuss* **128**, 55 (2005).

87. W. Q. Shi, Z. Q. Wang, S. X. Cui, X. Zhang and Z. S. Bo, *Macromolecules* **38**, 861 (2005).
88. S. Zou, I. Korczagin, M. A. Hempenius, H. Schönherr and G. J. Vancso, *Polymer* **47**, 2483 (2006).
89. E. S. Gil and S. M. Hudson *Prog. Polymer Sci.* **29**, 1173 (2004).
90. W. T. S. Huck, *Mater. Today* **11**, 24 (2008).
91. P. D. Boyer, *Ann. Rev. Biochem.* **66**, 717 (1997).
92. H. Y. Wang and G. Oster, *Nature* **396**, 279 (1998).
93. J. Howard, *Mechanics of Motor Proteins and the Cytoskeleton* (Sinauer Associates, Massachusetts, 2001).
94. R. Lipowsky and S. Klumpp, *Physica A* **352**, 53 (2005).
95. T. Atsumi, L. McCarter and Y. Imae, *Nature* **355**, 182 (1992).
96. W. Shi, M. I. Giannotti, X. Zhang, M. A. Hempenius, H. Schönherr and G. J. Vancso, *Angew. Chem. Int. Ed.* **46**, 8400 (2007).
97. H. J. Butt, *Macromol. Chem. Phys.* **207**, 573 (2006).
98. N. B. Holland, T. Hugel, G. Neuert, A. Cattani-Scholz, C. Renner, D. Oesterhelt, L. Moroder, M. Seitz and H. E. Gaub, *Macromolecules* **36**, 2015 (2003).
99. T. Hugel, N. B. Holland, A. Cattani, L. Moroder, M. Seitz and H. E. Gaub, *Science* **296**, 1103 (2002).
100. W. Q. Shi, S. Cui, C. Wang, L. Wang, X. Zhang, X. J. Wang and L. Wang, *Macromolecules* **37**, 1839 (2004).
101. S. Zou, Y. Ma, M. A. Hempenius, H. Schönherr and G. J. Vancso, *Langmuir* **20**, 6278 (2004).
102. H. G. Schild, *Prog. Polymer Sci.* **17**, 163 (1992).
103. Z. Ding, G. Chen and A. S. Hoffman, *Biocon. Chem.* **7**, 121 (1996).
104. F. Tanaka, T. Koga, H. Kojima and F. M. Winnik, *Macromolecules* **42**, 1321 (2009).
105. E. Kutnyanszky, A. Embrechts, M. A. Hempenius and G. J. Vancso, to be published.
106. S. K. Ahn, R. M. Kasi, S. C. Kim, N. Sharma and Y. Zhou, *Soft Matter* **4**, 1151 (2008).
107. F. Zhou and W. T. S. Huck, *Phys. Chem. Chem. Phys.* **8**, 3815 (2006).
108. J. Wang, M. I. Gibson, R. Barbey, S. J. Xiao and H. A. Klok, *Macromol. Rapid Comm.* **30**, 845 (2009).
109. X. Sui, S. Zapotoczny, E. M. Benetti, P. Schön and G. J. Vancso, *J. Mat. Chem.* **20**, 4981 (2010).
110. A. J. Parnell, S. J. Martin, C. C. Dang, M. Geoghegan, R. A. L. Jones, C. J. Crook, J. R. Howse and A. J. Ryan, *Polymer* **50**, 1005 (2009).
111. C. Ortiz and G. Hadziioannou, *Macromolecules* **32**, 780 (1999).
112. G. Strobl, in *The Physics of Polymers: Concepts for Understanding Their Structures and Behavior* (Springer, 2007).
113. H. B. Li, B. B. Liu, X. Zhang, C. X. Gao, J. C. Shen and G. T. Zou, *Langmuir* **15**, 2120 (1999).
114. E. Evans, *Ann. Rev. Biophys. Biomol. Struct.* **30**, 105 (2001).
115. G. Neuert, T. Hugel, R. R. Netz and H. E. Gaub, *Macromolecules* **39**, 789 (2006).

116. K. Kulbaba and I. Manners, *Macromol. Rapid Comm.* **22**, 711 (2001).
117. J. Skolnick and M. Fixman, *Macromolecules* **10**, 944 (1977).
118. T. Odijk, *Macromolecules* **12**, 688 (1979).
119. D. Foucher, B. Z. Tang and I. Manners, *J. Am. Chem. Soc.* **114**, 6246 (1992).
120. U. S. Schubert, G. R. Newkome and I. Manners, *Metal-Containing and Metallo-supramolecular Polymers and Materials* (eds.), (American Chemical Society, Washington, DC, 2006).
121. S. Zou, M. A. Hempenius, H. Schönherr and G. J. Vancso, *Macromol. Rapid Comm.* **27**, 103 (2006).

Marina I. Giannotti was born in Mar del Plata, Argentina, in 1975. She graduated in Chemistry (1999) at the University of Mar del Plata, where she also received her PhD in Materials Science (2004) with Prof. P. A. Oyanguren in the group of Prof. R. J. J. Williams (INTEMA/CONICET), working on polymerization-induced phase separation and toughening of thermosetting polymers. In 2004, she was awarded a Marie Curie European postdoctoral fellowship and she moved to the University of Twente, The Netherlands, where she worked with Prof. G. J. Vancso on nanotechnology with single macromolecules using AFM-based single molecule force spectroscopy. Since 2008, she has been a researcher at the Biomedical Research Networking Centre in Bioengineering, Biomaterials and Nanomedicine (CIBER-BBN), Spain, in the group of Prof. F. Sanz at the University of Barcelona. Dr Giannotti's interests comprise the fields of macromolecular nanotechnology and biomedical applications, and AFM-based nanomechanical studies.

Edit Kutnyánszky was born in Dombóvár, Hungary, in 1984. She received her M.Sc. in Chemistry (2008) at Eötvös Loránd University, Hungary. In 2008, she joined the group of Prof. Julius Vancso at the University of Twente, The Netherlands, as a PhD researcher. Her project aims at synthesis of stimulus-responsive polymer systems and their characterization with atomic force microscopy–based force spectroscopy.

G. Julius Vancso studied Physics at Eötvös Loránd University of Sciences in Budapest, and Materials Science at the Swiss Federal Institute of Technology (ETH) in Zurich. He completed his PhD thesis on electroactive polymers in 1982. He was a postdoctoral fellow and lecturer at the ETH, in its Polymer Institute (1983–88); Assistant and then Associate Professor with tenure at the University of Toronto, Canada, in the Department of Chemistry and the Department of Metallurgy and Materials Science (1988–1995); and has since 1995 been full Professor and Chair in Materials Science and Technology of Polymers of the University of Twente in the MESA+ Institute for Nanotechnology, in Enschede, The Netherlands. Professor Vancso's interests revolve around macromolecular nanotechnology, polymer material chemistry and stimulus-responsive macromolecules. He has authored or coauthored over 360 publications. He is currently a Visiting Principal Scientist at the Institute of Materials Research and Engineering of A*STAR in Singapore, and founding Senior Editor of *European Polymer Journal*'s Macromolecular Nanotechnology Section. He was elected Foreign Member of the Hungarian Academy of Sciences in 2010.

Probing Human Disease States Using Atomic Force Microscopy

Chapter

5

Ang Li*,† and Chwee Teck Lim*,†,‡,§

Abstract

Atomic force microscopy (AFM) has been in use for more than a decade and has become a routine "nano/micro-topography" imaging tool as well as a nanomechanical property characterization tool in a wide range of disciplines, including those of biomedical and disease research. Some disease states are accompanied by biomechanical property and morphological changes which occur at the nano- and microscales and can best be assessed with specific advanced techniques such as AFM. In this chapter, the technical advances and some recent applications of using AFM to study human diseases such as malaria and cancer will be reviewed. The outlook for the prospective technical developments in AFM that will facilitate further understanding of disease pathology will also be discussed.

*Division of Bioengineering and Department of Mechanical Engineering, National University of Singapore, 9 Engineering Drive 1, Singapore 117576.

†Singapore–MIT Alliance, National University of Singapore, 4 Engineering Drive 3, Singapore 117576.

‡Mechanobiology Institute, National University of Singapore, 5A, Engineering Drive 1, Singapore 117411.

§Email: ctlim@nus.edu.sg

5.1. AFM as an Imaging Tool for Biological Applications

5.1.1. *Basic and advanced imaging modes*

The basic idea of AFM imaging is to sense the surface topography through a flexible cantilever with a sharp tip mounted at the end which is directly interacting with the surface. During scanning, the tip moves over the surface in a raster fashion with the help of a piezoelectric scanner with subnanometer positioning resolution. The deflection change of the cantilever is measured by reflecting a laser beam from the back side of the cantilever that falls onto a quadrant photodiode detector. The images can be converted from the scanner z-position recorded from the feedback system which controls the deflection or the amplitude of the cantilever depending on the mode of imaging.

To better understand different imaging modes developed for AFM, the fundamental physics of tip–sample interactions at the atomic or molecular scale needs to be recalled. The force–distance profile between the tip and the sample at a close position is similar to the Lennard–Jones potential profile, as shown in Fig. 1. When a bare tip (which has no specific interaction with the surface) approaches the surface from a long distance, it senses an attractive force first due to van der Waals interaction, which increases to a maximum before the force becomes repulsive gradually due to Pauli repulsion. Distinguished by the regions of interaction that the tip falls within during scanning, there are three basic modes of imaging, namely contact mode, intermittent contact or tapping mode, and noncontact mode. In contact mode, the tip is in close contact with the sample surface where repulsive force dominates tip–sample interaction and the deflection change of the cantilever is used as the feedback signal. Contact mode is the most commonly used imaging technique for AFM but it exerts considerable friction and indentation forces on the surface, which limits its applications on some delicate samples, including biological samples. To overcome the drawbacks of contact mode, intermittent contact and noncontact mode were developed. In these two modes, the cantilever is oscillated near its resonance frequency and the amplitude change of the cantilever is monitored by the feedback system. In intermittent contact mode, the central position

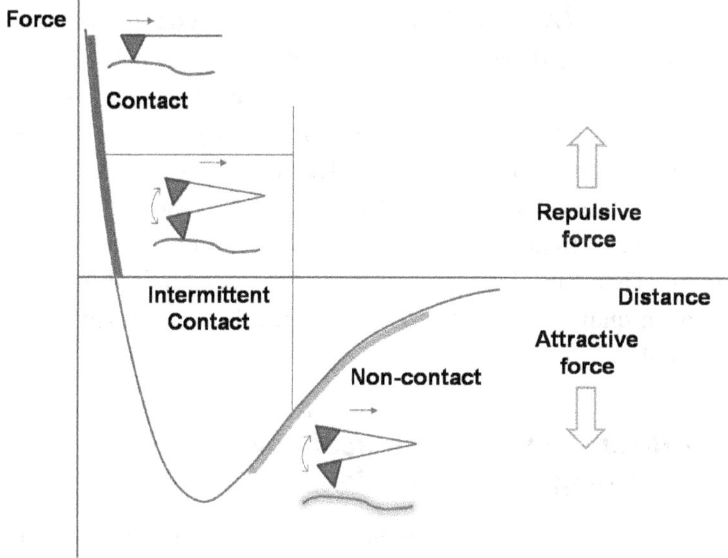

Fig. 1. Schematic plot of tip–sample interaction force as a function of distance and the working regions of different imaging modes of an AFM.

of the tip is maintained in the attraction region but the tip touches the surface when it extends in the oscillation cycle. Thus, this mode reduces the friction dramatically but there is not much reduction in the indentation force. In noncontact mode, the tip is operating fully in the attraction region but is always kept a few nanometers away from the surface. Noncontact mode can achieve near-zero friction and indentation, since the tip has little contact with the sample. However, the maintenance in the attraction region usually requires high surface attractive force and the surface needs to be fairly flat.

Different techniques have been developed to oscillate the cantilevers for tapping mode. In the traditional acoustic or piezodriven AC mode design, the cantilever chip is mounted in a cantilever holder equipped with a small piece of piezomaterial which oscillates the whole unit. Such passive oscillation design is simple and effective in air but the performance in the liquid environment is obscured by the damping effect from the liquid. To achieve better performance in wet conditions, two other types of active oscillation design have been developed, namely magnetic AC or MAC mode[1] and iDrive^{TM} AC mode (Asylum Research). In MAC mode, the back side of the cantilever is coated with a thin layer of magnetic materials and the oscillation of the cantilever is driven by an external AC magnetic field. By contrast, iDrive AC mode actuates the cantilever by means of running a small AC current through the cantilever legs in the presence of a constant magnetic field. These two improved tapping modes can achieve a small amplitude tapping effect and give rise to clearer tuning curves when working in liquid. Therefore the signal-to-noise ratio is significantly improved and the image quality is usually better, especially when imaging loosely attached small objects.

5.1.2. *Current state of technical developments for biological applications*

High resolution imaging of biological macromolecules has been achieved over the last 10 years of improvements of instrumentation as well as sample preparation techniques. Early molecular resolution images were obtained on immobilized DNA[2] or crystallized proteins that were originally prepared for electron microscopy and X-ray crystallography.[3] The molecular structures derived from AFM images of these proteins agree excellently with the results obtained from the other techniques and, more importantly, AFM imaging provides quantitative topography dimensions from which the protein volume or DNA length can be measured and correlated with their molecular weights or gene transcription levels.[4–7] Later developments of sample preparation techniques ushered in the AFM imaging of biomolecules in their native states under aqueous conditions. This was mainly achieved on supportive or nonsupportive lipid membrane from which conformational changes under different conditions can be studied *in situ*.[8–10] With this advantage, AFM can be used to monitor the structural changes of biomolecules simultaneously with biochemical treatments. For instance, opening and closing of ion channels in response to the stimulation of electrical potential, Ca^{2+} addition or pH changes have been clearly demonstrated at sub-nm resolution.[8,11,12]

Usually, contact mode is sufficient to produce high resolution images owing to the molecule at the very end of the probe serving as the tip interacting with the surface molecules. However, sometimes tapping mode is preferred due to its ability to produce "phase" images and thus provides extra contrast to the molecular structures. One problem associated with tapping mode performance in liquid is the damping effect from the surrounding liquid, which significantly reduces the quality (Q) factor of the oscillating cantilever. In addition to the previously mentioned MAC mode or iDrive mode, which can improve the imaging quality of tapping mode in liquid, two other complementary techniques can also help. One approach is to digitally enhance the effective Q-factor through an extra active feedback design.[13] This approach can increase the sensitivity of the cantilever to the external forces and thus be able to reduce the tapping force on the sample and minimize the scanning-induced deformation of the sample, which is a major source of distorted imaging. The other approach is bimodal AFM or DualAC™ mode (Asylum Research), in which the cantilever is excited simultaneously at the first two flexural modes instead of one for the traditional acoustic AC mode. In such a design, lower scanning force and higher compositional contrast, and thus better spatial resolution, can be achieved. The demonstration of the technique on imaging of individual antibodies has clearly shown the Fc (fragment crystallizable region) and Fab (fragment antigen binding) fragments in the second order phase imaging with submolecular resolution.[14]

In addition to the developments in high resolution imaging, AFM has also been adapted as a "live cell imaging" technique, which has revealed structural details of hydrated cell surfaces under near-physiological conditions with extraordinary resolution. Sample preparation is again one of the key factors for achieving successful results. In order to resist the lateral forces exerted by the scanning tip, usually the sample is required to be immobilized on a flat and rigid substrate. One straightforward immobilization method is to drop the sample solution onto a substrate and leave the sample to dry in a vacuum desiccator or under nitrogen blow. To prevent the collapse of cell structures due to the loss of water and surface tension during drying, chemical fixation of the cells using cross-linking agents such as glutaraldehyde or formaldehyde may be necessary. Both contact mode and tapping mode can be applied to image dried samples and the resolution can be fairly high because the sample is well immobilized and fixed after being dried. However, this approach is not physiologically preferred and may introduce artefact due to excessive cross-linking. Also, if the sample is dried in the culture solution, the salt crystal deposits from the solution will severely contaminate the sample surface.

In order to study cell structures in conditions that are as close to the natural state of the cells as possible, working in liquid is generally required. Adherent types of cells (endothelial cells, for example) can be directly grown on a substrate to achieve sufficient affinity and surface tension to withstand the applied scanning force.[15] However, this method is dependent on the cell line and, often, the adhesion is not strong enough for a short culture period. Thus, an appropriate coating of the substrate with adhesion molecules is a feasible approach, which offers the advantage of allowing the sample to be directly imaged under liquid without the need for air-drying. One of the most commonly used nonspecific adhesive molecules is poly-L-lysine. The polycationic nature of the molecule allows interaction with the anionic sites on the cells, resulting in enhanced adhesion to negatively charged mica or glass surface. Alternatively, extracts from extracellular matrix such as fibronectin and collagen can also enable cells to adhere to the substrates. In order to achieve specific and stronger affinity, another approach is to introduce cross-linking groups to facilitate chemical bonding between the sample and the surface. APTES and ANB-NOS are the most commonly used chemicals to silanize mica or glass surface and generate amine groups for further covalent binding with different reactive groups on the sample surface.[16]

AFM is ideal for examining surface structures with sufficient topographical contrast but lacks biological specificity, i.e. it has limited ability to distinguish different objects with similar topography features. By contrast, optical imaging methods enable spectroscopic discrimination of different biological organisms but are limited by the spatial resolution due to the light diffraction limit, which is about 200–300 nm. To address these limitations, recently developed integration of AFM and advanced optical microscopy techniques (confocal laser scanning microscopy, for instance) equips AFM with optical methods–based biochemical identification capability which opens up a new era of "BioAFM."[17,18] Besides, as with immunoelectron microscopy methods, AFM has also been developed into "immune AFM" which can identify specific surface molecules with the help of quantum dot markers.[19,20]

5.1.3. *AFM imaging study of malaria and Babesia-infected red blood cells*

5.1.3.1. *Malaria pathology: surface morphology as an indicator of the disease state and association with pathology*

Malaria is one of the most severe infectious diseases on earth. Malaria parasites are transmitted by *Anopheles* mosquitoes and target human red blood cells (RBCs). Once the parasite invades an RBC, it rapidly develops into intracellular states and remodels the host cell both structurally and functionally. Specific ultrastructures have been observed at the surface of infected red blood cells (IRBCs) after the parasites developed into mature stages. These surface structures are species-specific and are linked to disease pathology and virulence. For example, the most deadly species of human malaria, *Plasmodium falciparum*, develops surface protrusions called "knobs," which concentrate parasite origin adherent molecules that help the IRBCs to stick to the endothelial cells lining the blood vessels. This phenomenon is known as cytoadherence and is believed to promote the survival of parasites by keeping them from passing through the spleen, where they would be cleared due to their abnormal decreased deformability. However, this is also one of the major pathological factors associated with severe malaria, as the sticking infected cells would block the flow of normal red cells in the small capillaries, thus preventing them from transporting oxygen and nutrients to the vital organs. Therefore, a detailed understanding of the surface features of RBCs at the subcellular level is necessary.

5.1.3.2. *Methods and results*

To image normal and infected RBCs, a simple smear method has been developed to prepare unfixed structurally preserved samples.[21] Thin smears of healthy RBCs were made on a clean glass slide and dried in a vacuum desiccator. *Plasmodium falciparum* laboratory clone 3D7 was cultured routinely, following standard procedures.[22] Thin smears of malaria IRBCs were prepared in the same way as those of healthy RBCs except that the IRBCs were stained with DAPI or Giemsa staining reagents prior to smearing. Anderson and K strains of *Babesia bovis* which originated from an infected cow were maintained in culture and the thin smear samples were prepared in the same way as those for malaria IRBCs.

The instrument used for this study was a Dimension 3100 system equipped with a Nanoscope IIIA controller (Veeco Instrument). Standard tapping mode in air was applied to image the whole cell as well as the fine surface ultrastructures. The 3D surface plot and coloring effect of the AFM images shown here were processed using the SPIP software.

Dramatic changes on the surface of normal and malaria or *Babesia* IRBCs can be clearly seen in Fig. 2. The surface of normal RBCs exhibited a network structure which showed the spectrin network underneath the cell membrane [Fig. 2(B)]. This is mainly due to the drying-induced collapse of lipid membrane, which exposes the rigid cytoskeletal structures underneath. On the surface of malaria IRBCs, numerous knoblike protrusions measuring about tens of nm high and about a hundred nm in the base diameter were observed [Fig. 2(D)]. These knobs appear only at the late infection stages and cover the whole cell surface at a density of about 5–7/μm^2. On the surface of *Babesia*-infected cells, protrusions were also observed but they were randomly oriented with ridgelike shapes called "ridges" [Fig. 2(F)]. These ridges increase in number when the parasite matures within the host cell. More ridges were found on the surface of the virulent Anderson strain compared with the nonvirulent K strain, which may imply a correlation between these surface structures and disease severity.

5.1.3.3. *Discussion*

Surface-morphological changes are of pathological significance to these parasite infection diseases. The surface protrusions — knobs for malaria IRBCs and ridges for *Babesia*-infected cells — have been shown to be the aggregation sites of parasite origin adherent proteins which function as "bridges" anchoring infected cells to the endothelial cells lining the blood vessel wall. The quantitative morphological studies presented here can help to characterize these species-specific surface "fingerprints" and their correlations with developmental stages and disease virulence. Future studies with drug treatments will provide a reliable assessment method for different drug candidates. Also, our study has paved the way for further

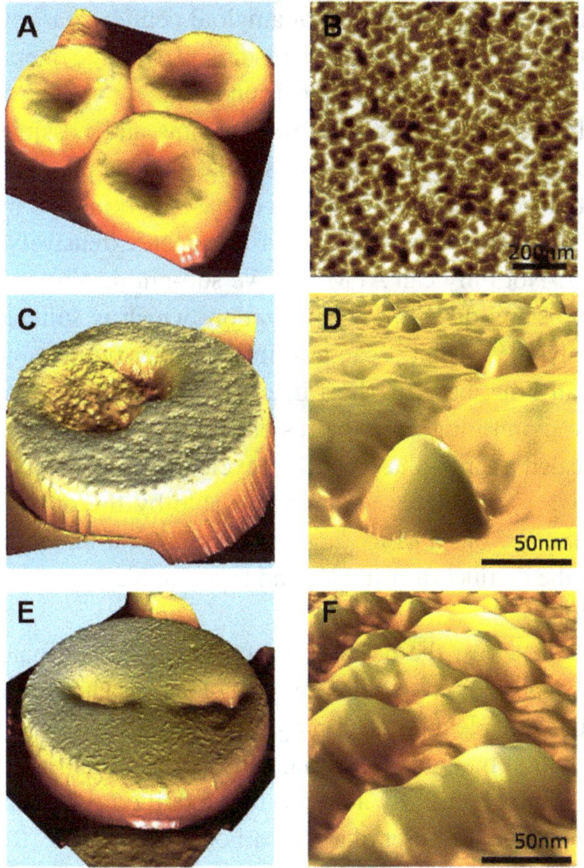

Fig. 2. AFM images of (A) normal RBCs and (B) their spectrin network surface structures compared with (C, D) knobby structures on the surface of malaria IRBCs and (E, F) ridges on the surface of *Babesia*-infected cells.

molecular level research on the structural and functional changes associated with malaria infection.

5.1.4. *AFM imaging study of other diseases*

The capability of AFM to produce high resolution images under robust conditions makes it a useful technique for studying molecular and subcellular level structural changes associated with disease states. At the molecular level, AFM imaging of aggregation of amyloid peptides has revealed their structural details, fibril-forming mechanisms and dynamic interactions with different antibodies under physiological conditions.[23–27] Such studies have provided valuable insights into the mechanisms and possible therapeutic strategies of a series of diseases associated with protein misfolding and aggregation, including Alzheimer's disease, Parkinson's diseases and diabetes. At the subcellular level, AFM has been extensively applied to image viruses. Not only can AFM achieve sub-nm resolution, as high as electron microscopy or X-ray crystallization, which is suitable for ultrastructural analysis,[28,29] but it can also image under liquid conditions and monitor in real time the virus–host interaction.[30–33] As an example, real time dynamic monitoring of retrovirus budding from the surface of living cells has revealed a three-phase budding cycle that follows two kinetically distinct pathways,[32] which was not achievable by the other techniques. In addition, the use of AFM to diagnose certain viral infections is possible owing to the distinct dimensions and topographic features of different species of viruses.[34]

Besides the virus, the bacterium represents another major group of pathogens that cause disease. AFM imaging of bacteria has been conducted for more than a decade and much progress has been achieved in terms of sample preparation methods. Bacterial cells can be immobilized for AFM scanning by being chemically attached to a modified substrate surface or physically trapped in the pores of a polycarbonate membrane. Recent applications of AFM imaging on the cellular structures and cell remodeling of bacteria include *E. coli*,[35] *Bacillus*,[36,37] *Clostridium*,[38] *Mycobacterium*,[39] *Myxococcus*[40] and others. With particular interest, AFM imaging can also help to gain insights into the mechanisms of antibiotics or antimicrobial peptide action on bacterial cells.[41–44]

At the cellular level, morphological changes are often associated with mammalian cells at disease states and AFM can also help to understand the pathophysiology and assess different therapeutic methods. For example, cisplatin-encapsulated liposomes have been studied in terms of their physical properties and stability as well as the kinetics of cell internalization and drug release using a combined optical and AFM imaging technique.[45] Moreover, AFM imaging can be conducted directly on tissues to obtain the molecular mechanism of certain diseases, such as cataract.[46,47]

In summary, the unique ability of AFM to image a variety of biological samples ranging from molecules to tissues in their natural aqueous environments with fairly high resolution makes it a particular powerful tool for studying diseases. However, topography imaging alone cannot provide comprehensive information, due to the lack of biological specificity. Thus, complementary methods such as optical and mechanical measurements have to be used together to distinguish topographically identical objects. In the next part of this chapter, the use of AFM as a nano/micro-mechanical measurement tool in the study of diseases will be discussed.

5.2. AFM as a Force-Sensing Tool (Nano- and Micromechanical Property Measurements Using AFM)

5.2.1. *Force measurement and property-mapping techniques*

The most basic means of measuring force in an AFM system is to move the cantilever in the vertical direction relative to the sample surface in an approach–contact–retract cycle while monitoring the deflection of the cantilever versus the tip position. This full cycle is recorded in the form of a force–distance (F–D) curve, from which all kinds of interaction information can be extracted. The approaching curve can be used to measure the elastic property of the sample and the retraction curve can be used to measure adhesion. Figure 3 schematically shows an example of a force curve obtained on a cell using a tip functionalized with binding receptors. When the tip comes into contact with the cell surface, the resultant deflection of the cantilever is not linearly related to the tip position (stage B in Fig. 3), as it is mainly due to the contact area change (an increasing contact area because of the tip geometry) and viscoelastic response from the cell. When the tip is retracted and moved away from the cell, the deflection of the cantilever first recovers and then deflects in the other direction due to the binding force induced between the tip-bound receptors and cell surface ligands. When the bending force of the AFM cantilever exceeds the binding force, the tip detaches and the cantilever is restored to its original position

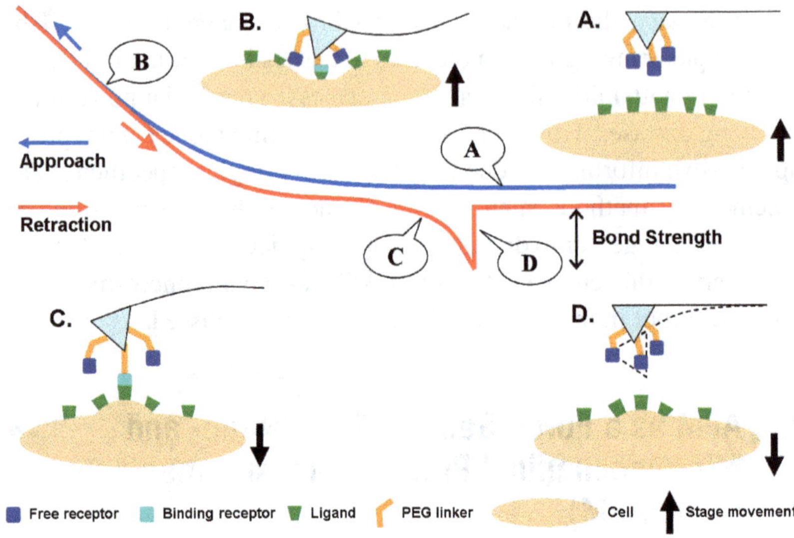

Fig. 3. Schematic of different stages of a force–displacement curve when an AFM tip functionalized with binding receptors first approaches (A), contacts (B), retracts (C) and finally detaches (D) from a cell surface.

and a peak — shown in stage D in Fig. 3 — is registered. This peak indicates the unbinding force needed to break the adhesion between the tip and the cell, and can be used to quantify the specific single molecular binding force. Thus, from one F–D curve, one can measure both the elastic/viscoelastic property of the sample and the adhesion force/energy between the tip and the cell.

Simple force curve mode only collects force information at a single point, whereas force volume mapping automatically collects series of force curves in an array of a specified area. Since the x–y position is also recorded in force volume mode, it enables the characterization of both topography and surface interaction force which can be converted to certain surface properties. Thus, in principle, force volume mode can provide decoupled topography and surface property information of the sample. Nearly all types of surface properties can be investigated in force volume mode but certain tip functionalizations may be required.

In terms of biological applications, adhesion mapping and elasticity mapping are the most common uses of force volume mode. By functionalizing the tip with specific ligands, single molecular interaction force measurements and molecular recognition on living cells can be achieved.[48] Also, force volume mode can be used to investigate the elastic properties of the sample or produce a microelasticity map that shows local variations in surface stiffness by measuring the force required to indent or deform the surface. Elasticity measurements of biological cells in response to rearrangements of cytoskeletal elements and other cellular or environmental changes have been extensively studied.[49,50]

5.2.2. *Nanoindentation of cancer cells as an example*

5.2.2.1. *Background*

Cancer is one of the leading causes of death. It arises due to the rapid and uncontrolled division of abnormal cells. Some cancers cause cells to change their physical properties, such as their deformability and adhesion, and this may have resulted in the likelihood of these cancerous cells migrating to other parts of the body through a process called "metastasis."[51] Such pathophysiological changes may be related to changes in the cytoskeletal structures. For example, study of breast cancer cell lines suggests that the benign MCF-10A cell possesses a cytoskeleton that has an ordered and more rigid structure than the malignant MCF-7 cell, which possesses a more disorganized and compliant cytoskeletal structure.[52] Such cytoskeletal alteration results in changes to the overall mechanical properties of the cell and hence may have given rise to their pathophysiology.[53]

5.2.2.2. *Method and results*

The elastic moduli of two types of human breast epithelial cells — benign MCF-10A and malignant MCF-7 cells — were probed using a multimode AFM equipped with a picoforce module (Veeco Instruments). A 4.5-μm-diameter polystyrene bead attached to a 0.01 N/m silicon nitride cantilever (NovaScan, USA) was used as the indenter. The spherical indenter can facilitate the modeling and assessment of cell elasticity through a well-defined contact geometry.

Indentation was carried out at the center of the cell using different loading rates from 0.03 to 1 Hz. An indentation force of 200 pN was applied

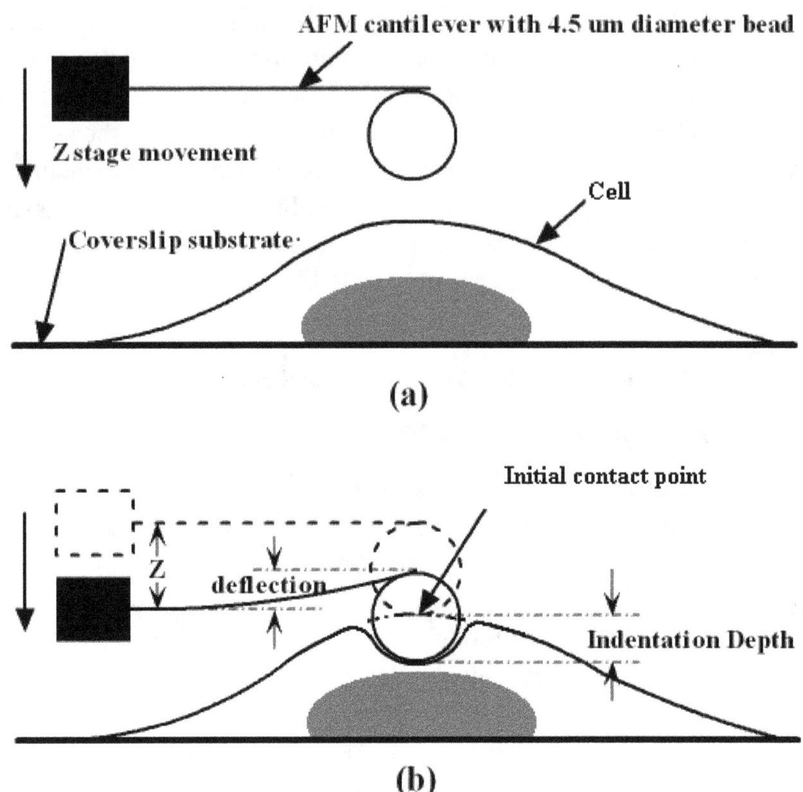

Fig. 4. The schematic shows the experimental setup and geometry changes (a) before and (b) after tip-cell contact. (*Reprinted with permission from Ref. 52.*)

to ensure a small deformation of the cell and minimize substrate effects. The Young's modulus was subsequently determined using Hertz's contact model.

By changing the loading rates, indentation curves showing different cell stiffnesses were clearly seen [Fig. 5(a)]. Apparently, fast indentation gave rise to stiffer results as compared to slow indentation, due to the viscoelastic response of the cells. Thus, cells appear to be stiffer at a higher loading rate.

Also, the comparison of benign and maglinant cells at physiologic temperature showed significant difference on the apparent Young's modulus at a given loading rate [Fig. 5(b)]. It is evident that the benign MCF-10A

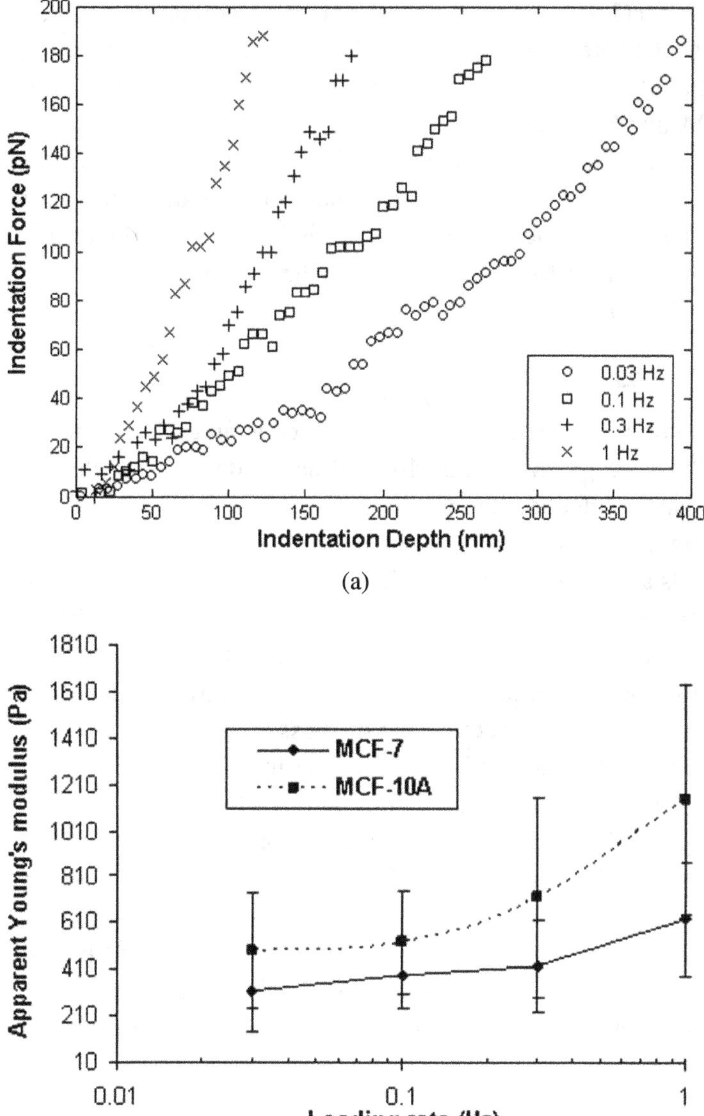

Fig. 5. (a) Indentation curve and (b) the calculated apparent Young's modulus at different loading rates. (*Reprinted with permission from Ref. 52.*)

appears stiffer (1.4–1.8 times higher in Young's modulus) than the malignant MCF-7 at the same loading rate, and also higher values of the apparent Young modulus were associated with an increased loading rate for both MCF-10A and MCF-7.

AFM imaging and confocal imaging were also applied to image the cellular structures of benign and maglinant breast cancer cells. It appears that the cytoskeletal actin structure of MCF-7 is less organized and not as fibrous as that of MCF-10A; this may have contributed to the reduced stiffness of MCF-7 cells.

5.2.2.3. *Discussion*

The correlation between deformability of cells and disease states is well known.[51] Developments of AFM-based nanoindentation techniques have enabled cancer detection and mechanism study at the single cell level with more certainty. In this study, we clearly show that the malignant breast cancer cells are more deformable than the benign cells, due to the changes at the subcellular cytoskeleton level. Also, the results bring to our attention the loading rate effect of obtaining different values of Young's modulus under different rates of indentation. Thus, the relative "apparent" Young's modulus was used to characterize the stiffness of cells instead of the absolute value. This is also a general consideration in elasticity measurements of single cells using AFM, due to the limitation of the assumptions of Hertz's contact model. In a recent study conducted on clinical samples, a similar magnitude of reduction in terms of Young's modulus in metastatic cancer cells to that of benign cells was observed and this mechanical property was suggested as a biomarker for cancer detection.[54] It should be noted that the normal cells in this study were collected from the same site (the lung, chest and abdomen of patients) as the cancerous cells, thus strengthening the conclusions.

5.2.3. *General applications in disease studies using AFM-based force spectroscopy and nanoindentation techniques*

The unique capabilities of AFM to systematically probe different types of surface physical properties of biological materials have extended the applications of AFM from topography imaging to a broad range of functional imaging. AFM-based single-molecule force spectroscopy and chemical force microscopy can specifically detect intra- and intermolecular interaction forces and certain surface chemical properties. By applying these techniques in disease research, all kinds of biological and microbiological questions have been addressed, including the surface adhesion properties of bacterial cells,[55–57] antibiotics–bacteria interaction,[58] virus–cell interaction,[59,60] nanomechanical properties of amyloid aggregates,[61] protein misfolding[62] and pathological ligand–receptor interactions.[63]

Many pathological processes can affect cell structures and, as a result, alter the biomechanical properties of these cells and tissues.[51] The AFM-based nanoindentation method can detect local elastic and viscoelastic properties of cells in their native conditions and thus is a promising method for diagnosing different pathologies. For example, in the case of hemolytic anemia, the Young's modulus of the RBCs collected from patients was 2–3 times higher than for normal RBCs.[64] In a tissue level study, chondrocytes isolated from osteoarthritic human cartilage exhibited an elevated viscoelastic modulus as compared to cells from normal tissues.[65] Aging was also found to have a significant impact on the stiffness of cardiac myocytes.[66] As a consequence, AFM has been suggested as an early detection tool for aging and osteoarthritis[67] as indentation-type AFM can monitor early damage in the cartilage of osteoarthritic patients before morphological changes can be observed using other diagnostic methods.

5.3. Outlook and Insights

AFM has become a versatile tool capable of high resolution imaging, live cell imaging, dynamic imaging, force spectroscopy and nanoindentation. The technique is under such rapid development that many promising features can be expected; among them, high-speed imaging and functional imaging are briefly mentioned here.

Current AFM imaging techniques suffer from temporal resolution, i.e. relatively slow scanning speed, mainly hindered by mechanical constraints of resonance of the scanner and response time of the cantilever. To achieve video rate scanning speed, small cantilevers with higher resonance frequency and faster response to mechanical interaction have been devised, and millisecond temporal resolution and piconewton force sensitivity have been achieved.[68,69] These technical advances will enable real-time monitoring of the dynamics of many biological processes.

Further developments in functional imaging beyond topography will also be of particular interest to disease study. Some current technical advances have shown exciting progress in simultaneous functional imaging of different surface features such as chemomechanical mapping of surface receptors and quantitative phase imaging of material hardness.[70,71] Together with the developments in the integration of AFM with other techniques, a comprehensive picture of the structure–property–function relationship of the biological specimens and their changes under different disease states will be clearly drawn and effective therapeutic strategies can then be accordingly developed.

Acknowledgments

We acknowledge support from the NUS Life Science Institute, Singapore–MIT Alliance (SMA), Singapore–MIT Alliance for Research and Technology (SMART), and Global Enterprise for Micro Mechanics & Molecular Medicine (GEM4).

References

1. G. Ge, D. Han, D. Lin, W. Chu, Y. Sun, L. Jiang, W. Ma and C. Wang, *Ultramicroscopy* **107**, 299 (2007).
2. D. A. Erie, G. Yang, H. C. Schultz and C. Bustamante, *Science* **266**, 1562 (1994).
3. F. A. Schabert, C. Henn and A. Engel, *Science* **268**, 92 (1995).
4. J. Lärmer, S. W. Schneider, T. Danker, A. Schwab and H. Oberleithner, *Pflügers Arch./Eur. J. Physiol.* **434**, 254 (1997).
5. J. Reed, B. Mishra, B. Pittenger, S. Magonov, J. Troke, M. A. Teitell and J. K. Gimzewski, *Nanotechnology* **18**, 044032 (2007).
6. H. Schillers, *Pflügers Arch./Eur. J. Physiol.* **456**, 163 (2008).

7. Y. Yang, H. Wang and D. A. Erie, *Methods*, **29**, 175 (2003).
8. A. Engel and D. J. Muller, *Nat. Struct. Mol. Biol.* **7**, 715 (2000).
9. R. P. Goncalves, G. Agnus, P. Sens, C. Houssin, B. Bartenlian and S. Scheuring, *Nat. Methods* **3**, 1007 (2006).
10. F. Orsini, M. Santacroce, P. Arosio, M. Castagna, C. Lenardi, G. Poletti and F. Sacchi, *Eur. Biophys. J.* **38**, 903 (2009).
11. D. J. Muller, G. M. Hand, A. Engel and G. E. Sosinsky, *EMBO J.* **21**, 3598 (2002).
12. J. Yu, C. A. Bippes, G. M. Hand, D. J. Muller and G. E. Sosinsky, *J. Biol. Chem.* **282**, 8895 (2007).
13. L. Chen, X. Yu and D. Wang, *Ultramicroscopy* **107**, 275 (2007).
14. N. F. Martinez, J. R. Lozano, E. T. Herruzo, F. Garcia, C. Richter, T. Sulzbach and R. Garcia, *Nanotechnology* **19**, 384011 (2008).
15. R. Matzke, K. Jacobson and M. Radmacher, *Nat. Cell. Biol.* **3**, 607 (2001).
16. F. Liu, J. Burgess, H. Mizukami and A. Ostafin, *Cell Biochem. Biophys.* **38**, 251 (2003).
17. A. Trache and G. A. Meininger, *J. Biomed. Opt.* **10**, 064023 (2005).
18. J. Madl, S. Rhode, H. Stangl, H. Stockinger, P. Hinterdorfer, G. J. Schutz and G. Kada, *Ultramicroscopy* **106**, 645 (2006).
19. M. Murakoshi, K. Iida, S. Kumano and H. Wada, *Pflugers Arch.* **457**, 885 (2009).
20. H. Schillers, *Pflugers Arch.* **456**, 163 (2008).
21. A. Li, A. H. Mansoor, K. S. Tan and C. T. Lim, *J. Microbiol. Methods* **66**, 434 (2006).
22. W. Trager and J. B. Jensen, *Science* **193**, 673 (1976).
23. J. Legleiter, D. L. Czilli, B. Gitter, R. B. DeMattos, D. M. Holtzman and T. Kowalewski, *J. Mol. Biol.* **335**, 997 (2004).
24. F. Moreno-Herrero, M. Perez, A. M. Baro and J. Avila, *Biophys. J.* **86**, 517 (2004).
25. F. Moreno-Herrero, J. M. Valpuesta, M. Perez, J. Colchero, A. M. Baro, J. Avila and E. Montejo De Garcini, *J. Alzheimer's Dis.* **3**, 443 (2001).
26. W. S. Gosal, S. L. Myers, S. E. Radford and N. H. Thomson, *Protein. Pept. Lett.* **13**, 261 (2006).
27. C. Goldsbury, J. Kistler, U. Aebi, T. Arvinte and G. J. Cooper, *J. Mol. Biol.* **285**, 33 (1999).
28. M. Plomp, M. K. Rice, E. K. Wagner, A. McPherson and A. J. Malkin, *Am. J. Pathol.* **160**, 1959 (2002).
29. S. Lin, C. K. Lee, S. Y. Lee, C. L. Kao, C. W. Lin, A. B. Wang, S. M. Hsu and L. S. Huang, *Cell Microbiol.* **7**, 1763 (2005).
30. M. Moloney, L. McDonnell and H. O'Shea, *Ultramicroscopy* **100**, 163 (2004).
31. Y. G. Kuznetsov, J. G. Victoria, W. E. Robinson, Jr. and A. McPherson, *J. Virol.* **77**, 11896 (2003).
32. M. Gladnikoff and I. Rousso, *Biophys. J.* **94**, 320 (2008).
33. J. W. Lee and M. L. Ng, *J. Nanobiotechnol.* **2**, 6 (2004).
34. E. V. Dubrovin, Y. F. Drygin, V. K. Novikov and I. V. Yaminsky, *Nanomedicine* **3**, 128 (2007).

35. M. A. Beckmann, S. Venkataraman, M. J. Doktycz, J. P. Nataro, C. J. Sullivan, J. L. Morrell-Falvey and D. P. Allison, *Ultramicroscopy* **106**, 695 (2006).
36. V. G. Chada, E. A. Sanstad, R. Wang and A. Driks, *J. Bacteriol.* **185**, 6255 (2003).
37. M. Plomp, T. J. Leighton, K. E. Wheeler, H. D. Hill and A. J. Malkin, *Proc. Natl. Acad. Sci. U.S.A.* **104**, 9644 (2007).
38. M. Plomp, J. M. McCaffery, I. Cheong, X. Huang, C. Bettegowda, K. W. Kinzler, S. Zhou, B. Vogelstein and A. J. Malkin, *J. Bacteriol.* **189**, 6457 (2007).
39. D. Alsteens, C. Verbelen, E. Dague, D. Raze, A. R. Baulard and Y. F. Dufrene, *Pflugers Arch.* **456**, 117 (2008).
40. A. E. Pelling, Y. Li, W. Shi and J. K. Gimzewski, *Proc. Natl. Acad. Sci. U.S.A.* **102**, 6484 (2005).
41. A. Li, P. Y. Lee, B. Ho, J. L. Ding and C. T. Lim, *Biochim. Biophys. Acta.* **1768**, 411 (2007).
42. P. C. Braga and D. Ricci, in *Atomic Force Microscopy* p. 179 (2004).
43. L. Yang, K. Wang, W. Tan, X. He, R. Jin, J. Li and H. Li, *Anal. Chem.* **78**, 7341 (2006).
44. C. Verbelen, V. Dupres, F. D. Menozzi, D. Raze, A. R. Baulard, P. Hols and Y. F. Dufrene, *FEMS Microbiol. Lett.* **264**, 192 (2006).
45. S. Ramachandran, A. P. Quist, S. Kumar and R. Lal, *Langmuir* **22**, 8156 (2006).
46. N. Buzhynskyy, J. F. Girmens, W. Faigle and S. Scheuring, *J. Mol. Biol.* **374**, 162 (2007).
47. A. Antunes, F. V. Gozzo, M. Nakamura, A. M. Safatle, S. L. Morelhao, H. E. Toma and P. S. Barros, *Micron* **38**, 286 (2007).
48. C. M. Franz, A. Taubenberger, P.-H. Puech and D. J. Muller, *Sci. STKE* 15 (2007).
49. A. H. E, W. F. Heinz, M. D. Antonik, N. P. D'Costa, S. Nageswaran, C. A. Schoenenberger and J. H. Hoh, *Biophys. J.* **74**, 1564 (1998).
50. K. D. Costa, *Dis. Markers* **19**, 139 (2003).
51. G. Y. Lee and C. T. Lim, *Trends Biotechnol.* **25**, 111 (2007).
52. Q. S. Li, G. Y. Lee, C. N. Ong and C. T. Lim, *Biochem. Biophys. Res. Commun.* **374**, 609 (2008).
53. B. Lincoln, H. M. Erickson, S. Schinkinger, F. Wottawah, D. Mitchell, S. Ulvick, C. Bilby and J. Guck, *Cytometry A* **59**, 203 (2004).
54. S. E. Cross, Y. S. Jin, J. Rao and J. K. Gimzewski, *Nat. Nanotechnol.* **2**, 780 (2007).
55. V. Dupres, F. D. Menozzi, C. Locht, B. H. Clare, N. L. Abbott, S. Cuenot, C. Bompard, D. Raze and Y. F. Dufrene, *Nat. Methods* **3**, 515 (2005).
56. R. J. T. Emerson, T. S. Bergstrom, Y. Liu, E. R. Soto, C. A. Brown, W. G. McGimpsey and T. A. Camesano, *Langmuir* **22**, 11311 (2006).
57. E. Dague, D. Alsteens, J. P. Latge and Y. F. Dufrene, *Biophys. J.* **94**, 656 (2008).
58. Y. Gilbert, M. Deghorain, L. Wang, B. Xu, P. D. Pollheimer, H. J. Gruber, J. Errington, B. Hallet, X. Haulot, C. Verbelen, P. Hols and Y. F. Dufrene, *Nano Lett.* **7**, 796 (2007).
59. A. Negishi, J. Chen, D. M. McCarty, R. J. Samulski, J. Liu and R. Superfine, *Glycobiology* **14**, 969 (2004).

60. M. I. Chang, P. Panorchan, T. M. Dobrowsky, Y. Tseng and D. Wirtz, *J. Virol.* **79**, 14748 (2005).
61. D. N. Ganchev, N. J. Cobb, K. Surewicz and W. K. Surewicz, *Biophys. J.* **95**, 2909 (2008).
62. J. Yu, S. Malkova and Y. L. Lyubchenko, *J. Mol. Biol.* **384**, 992 (2008).
63. X. Yan, Z. Liu and Y. Chen, *Acta Biochim. Biophys. Sin. (Shanghai)* **41**, 263 (2009).
64. I. Dulinska, M. Targosz, W. Strojny, M. Lekka, P. Czuba, W. Balwierz and M. Szymonski, *J. Biochem. Biophys. Methods.* **66**, 1 (2006).
65. F. Guilak, *Biorheology* **37**, 27 (2000).
66. S. C. Lieber, N. Aubry, J. Pain, G. Diaz, S. J. Kim and S. F. Vatner, *Am. J. Physiol. Heart Circ. Physiol.* **287**, H645 (2004).
67. M. Stolz, R. Gottardi, R. Raiteri, S. Miot, I. Martin, R. Imer, U. Staufer, A. Raducanu, M. Duggelin, W. Baschong, A. U. Daniels, N. F. Friederich, A. Aszodi and U. Aebi, *Nat. Nanotechnol.* **4**, 186 (2009).
68. T. Ando, N. Kodera, E. Takai, D. Maruyama, K. Saito and A. Toda, *Proc. Natl. Acad. Sci. U.S.A.* **98**, 12468 (2001).
69. M. B. Viani, L. I. Pietrasanta, J. B. Thompson, A. Chand, I. C. Gebeshuber, J. H. Kindt, M. Richter, H. G. Hansma and P. K. Hansma, *Nat. Struct. Biol.* **7**, 644 (2000).
70. S. Lee, J. Mandic and K. J. Van Vliet, *Proc. Natl. Acad. Sci. U.S.A.* **104**, 9609 (2007).
71. O. Sahin and N. Erina, *Nanotechnology* **19**, 445717 (2008).

Ang Li is a research fellow at the Singapore–MIT Alliance for Research and Technology (SMART) Centre, Singapore. He obtained his BSc from Fudan University in 2003 and his PhD from the National University of Singapore in 2008. His research interest is the applications of atomic force microscopy in cellular and molecular biomechanics. Dr Li has authored or coauthored over 10 scientific publications and delivered 8 scientific talks at international conferences and academic institutions. Some of his AFM images have been chosen to appear on journal and book covers. He has also won several scientific awards, including first prize of the SPMAGE09 contest, an international conference travel award, best presentation awards and best poster awards at international and local conferences.

Chwee Teck Lim is a Professor in the Division of Bioengineering and Department of Mechanical Engineering at the National University of Singapore. He is also a Principal Investigator at the university's Mechanobiology Institute. He heads the Nano Biomechanics Lab, which conducts research in cell and molecular biomechanics and mechanobiology, with a focus on human diseases such as malaria and cancer. Dr Lim has authored or coauthored more than 160 scientific publications and delivered more than 130 plenary/keynote/invited talks. He is currently on the editorial boards of eight international journals. He has also won several research awards and honors, including the highly cited author award, Best Paper awards, Best Poster awards and the Young Investigator Award at international conferences. His research was cited by the MIT's magazine *Technology Review* as one of the top 10 emerging technologies of 2006 that would "have a significant impact on business, medicine or culture."

Conducting Atomic Force Microscopy in Liquids

Chapter 6

Nitya Nand Gosvami* and Sean J. O'Shea†

Abstract

Liquids confined in a narrow gap between two surfaces can form ordered layers, which may lead to oscillatory-type solvation forces, which can be measured using atomic force microscopy (AFM). However, it remains experimentally challenging to study the detailed mechanics of a single AFM contact, and in this regard conducting AFM (C-AFM) can be used to reveal subtle changes in the contact junction which are not observed in standard force measurements. Furthermore, the ability to measure conductivity allows an evaluation of the tip–sample contact area, a difficult problem in AFM.

In this chapter, we present simultaneous force and current measurements as an AFM tip approaches a graphite surface in liquid and squeezes out the confined liquid. Experiments were performed on linear and branched alkanes. Solvation layering occurs for both types of liquid, but marked differences in the squeeze-out mechanics are observed, depending on whether the monolayer is ordered or disordered. It is found that continuum elastic models are well suited for describing the squeeze-out mechanism of ordered, solid-like materials. However, when the confined molecules are disordered, the data are qualitatively very different. Simple elastic models cannot be applied anymore and comparison with recent simulations suggests that some liquid molecules remain trapped within the junction.

*INM–Libniz Institute for New Materials, Campus D2 2, D-66123 Saarbruecken, Germany.
†Institute of Materials Research and Engineering (IMRE), Agency for Science, Technology and Research (A*Star), 3 Research Link, Singapore 117602.

6.1. Introduction

Understanding the interaction between two surfaces is of immense importance for various scientific and engineering problems.[1] Examples are friction, adhesion, wear, lubrication, deformation, colloidal suspensions and biological interactions.[1–3] The contact mechanics, both static and dynamic behavior, at the nanometer or single asperity level, underpins the complex surface interactions observed at the macroscopic scale. This is due to the fact that the contact between engineering surfaces is dominated by asperities.[3] The real contact area between two macroscopic bodies is the sum of the individual asperity contacts and, except for molecularly smooth surfaces, is always much smaller than the apparent contact area.

The real contact area is thus a key parameter and is required for calculations of various contact mechanics parameters. Several theoretical models have been developed for the mechanics of single asperity contacts. The first work by Heinrich Hertz[4] was later refined by Derjaguin, Muller and Toporov (DMT)[5] and Johnson, Kendall and Roberts (JKR)[6] to allow calculation of indentation depth, surface deformation, stresses, etc. for different cases of surface elasticity and adhesion. Furthermore, experimental techniques such as the surface force apparatus (SFA)[7] and the atomic force microscope (AFM)[8] now enable experiments to be performed with single asperity geometry, thus allowing theoretical models to be directly tested. AFM measurements, in particular, can be taken on a wide variety of different surfaces and in different mediums using tips with a typical radius of ~20 nm, i.e. AFM permits realistic, nanoscale contacts to be established and studied.

The problem of understanding interactions between two surfaces can become even more complex in the presence of an intervening liquid. The theoretical foundation of force interactions between two surfaces in a liquid medium was laid by Derjaguin, Landau, Verwey and Overbeek, and is known as the DLVO theory.[9,10] The theory explains interactions between the surfaces by taking into account two opposing long range interaction forces; the attractive van der Waals force and the repulsive double layer force,[11] which is electrostatic in origin. Later theoretical study[12] and experimental work using the SFA[7] showed the existence of oscillatory-type, short range forces in liquids when the separation between the surfaces approached a few molecular diameters. Such forces could not be explained by the DLVO theory and they result from a completely new phenomenon, namely the formation of liquid into discrete layers near surfaces. The forces corresponding to the formation of the layers are termed "solvation forces." Further complexity in force interactions in a liquid arises due to the fact that the intervening liquid itself can be complex (e.g. multicomponent mixtures), amphiphilic or polymeric. Also, the confining walls are not necessarily ideally smooth and can be amorphous, crystalline, rough, crystallographically aligned or misaligned, rigid or soft, and with varied surface chemistry.[2]

Solvation forces were first measured at the nanometer scale by AFM on a graphite surface, immersed in OMCTS and dodecanol.[13,14] Initial experiments were performed in static mode. Later dynamic mode measurements were performed showing that such techniques provide a more sensitive measurement.[15–17] A variety of liquids have subsequently been investigated using AFM.[18–20]

A particular interest for our discussion is to understand the effect of molecular geometry on solvation forces. For example, Lim *et al.*[21] reported the first experimental observation of solvation layering in a heavily branched alkane (squalane) using sample modulation AFM. This observation contrasted with early SFA results which did not reveal any oscillatory force for several branched alkanes, including squalane.[22] Granick *et al.* revisited the problem using SFA and showed that oscillatory forces can exist for squalane near mica, and suggested that the surface preparation method for SFA could have dramatic effects on the force measurements.[23] More recent experiments by Israelachvili *et al.* on squalane again showed only a monotonic force variation with no oscillations.[24] Thus, force experiments performed on branched alkanes remain controversial.[24,25] Apart from force measurements using SFA and AFM, recent experiments employing neutron and helium atom scattering[26,27] and X-ray reflectivity measurements[28] have indicated strong layering of squalane close to various solid surfaces, with squalane molecules lying parallel to the surface. Simulation has also shown that adsorbed squalane molecules can form ordered structures on a graphite surface.[29]

Quantitative interpretation of single asperity measurements requires knowledge of the real contact area between the two approaching surfaces, so that an estimate of the pressure or stresses involved in the systems can be made. Three approaches to obtaining direct information of the nanoscale contact area in AFM are to measure the friction,[30] lateral stiffness[31,32] or current from the tip apex. Lantz *et al.*[33] showed that the variation of friction, contact stiffness and current with load is proportional to the contact area found by continuum mechanics theory (specifically, the Maugis–Dugdale model). These results and others[34] provide strong evidence that the AFM tip can be treated as a single asperity contact, although considerable care is required in interpreting calculated parameters because of uncertainties at nanometer length scales, such as local roughness.[35] A good example of such uncertainty is the use of electrical measurements to obtain the *absolute* value of the contact area. This remains a serious challenge, due to limited information regarding nanoscale electrical properties of materials; for example, the mean free path of electrons is not known with great accuracy.[33] Also, for conduction across a confined organic film, as of interest in lubrication, the current variation is extremely sensitive to even subangstrom changes in separation between the AFM tip and the substrate.

Conducting AFM and continuum models are used throughout this chapter to interpret change in the tip–sample contact under varying loads. The alkanes used in this study form solvation layers on graphite. The layer closest to the graphite substrate is strongly bound, and can be regarded as an adsorbed monolayer. Pressure must be applied to squeeze out the molecular layers and bring the AFM tip and graphite surfaces together. Thus, the alkane + graphite system provides a model single asperity system for problems in boundary lubrication and point contact mechanics in liquid. Specifically, experimental data are presented that are related to the following problems: How is confined liquid squeezed out of a nanometer-sized gap? What are the effects of molecular branching and the fluidity of the confined liquid on the squeeze-out behavior? To what extent can the conducting AFM technique be used to study the conductivity and related mechanical behavior of confined fluids? What are the pressures involved? These are questions of scientific and technical interest in tribology and, as shown below, conducting AFM can elucidate such issues by revealing subtle changes in the contact junction which are not observed in standard force measurement.

6.2. Introduction to Conducting Atomic Force Microscopy (C-AFM)

Figure 1 shows a simple schematic diagram of conducting atomic force microscopy (C-AFM) used in this work. The commercial setup (Molecular Imaging Corp., USA) allows simultaneous measurement of tip–sample interaction forces as well as current flow between a conducting substrate and a metal (Au or Pt)–coated AFM tip. It is important that the tip apex maintains a conducting coating, because mechanical and electrical stress can wear the deposited material. The electrical integrity of the tip apex can be tested by ensuring that current can be measured at the adhesive minimum of the force curve.[36]

For C-AFM, the conducting tip is connected to the virtual ground of a current-to-voltage-amplifier (Keithley model 6485) with variable

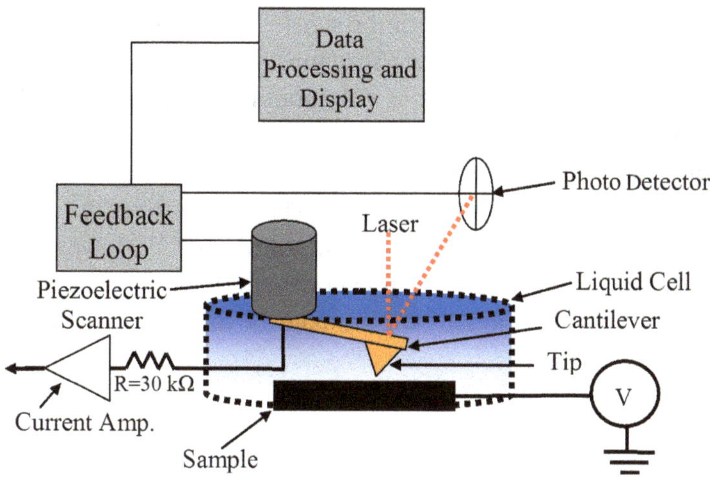

Fig. 1. Schematic of a conducting atomic force microscope (C-AFM). The bias voltage (V) is supplied by the AFM control electronics. The output from the current amplifier is read by an additional input channel of the AFM controller.

sensitivity of 1 V nA^{-1} to 1 V mA^{-1} and corresponding typical RMS noise levels of 20 fA to 100 nA. A 30 kΩ resistor is connected in series to limit the current. The sample is isolated from ground and connected to a bias voltage provided by the control electronics. STM experiments can be performed using the same setup. The measurements can be done either in air or in a liquid environment. A liquid environment provides much less contamination and eliminates capillary forces, allowing better control of tip–sample forces and improved imaging of soft materials such as self-assembled monolayers, or biological systems. To perform AFM experiments in liquids, a Teflon liquid cell is mounted on top of the sample. The cantilever is fully immersed inside the liquid cell.

6.3. Analysis of C-AFM Data

SFA and AFM data show good agreement with continuum theories developed for elastic single asperity contacts.[2,34] Plastic deformation can also occur in AFM experiments but the underlying analysis is semiempirical for nanoscale contacts. The continuum models described below assume a sphere-on-flat geometry, the deformations to be purely elastic, the materials to be isotropic and the elastic properties of the materials to remain unchanged under load. The atomic structure of the materials is not taken into account and the contact radius (a) should be small compared to the radius of the contacting sphere.

A general model, the Maugis–Dugdale (MD) model, was developed in 1992[37] and it reproduces the Hertz, DMT and JKR results as special cases. The MD model can be used to fit AFM data.[33] However, analysis using the full MD approach can become overly complicated. Carpick *et al.* simplified the MD model[38] to eliminate the mathematical difficulties, to give the contact radius

$$\frac{a}{a_{0(\alpha)}} = \left(\frac{\alpha + \sqrt{1 - F_a/F_c}}{1 + \alpha} \right)^{2/3}, \tag{1}$$

where $a_{0(\alpha)}$ is the contact radius at zero applied load, F_a the applied force and F_c the adhesion or pull-off force. The parameter $\alpha = 1$ corresponds to the JKR model and $\alpha = 0$ gives the DMT model. α is related to the original Maugis parameter λ by an empirical expression:

$$\lambda = -0.924 \cdot \ln(1 - 1.02\alpha). \tag{2}$$

In the work below, current versus force data obtained by employing C-AFM are fitted using Eq. (1) to provide values of the parameters $a_{0(\alpha)}$ and α. The appropriate continuum model (DMT, Hertz, JKR) can be directly recognized based on the numerical value of α.

The use of C-AFM also requires an understanding of how current flows at nanometer length scales. The description of charge transport between two metallic electrodes, either in physical contact or separated by a nanoscale gap (such as a monolayer of organic molecules), differs markedly from conduction at macroscopic length scales. For a macroscopic junction area, the resistance is defined by Ohm's law, where current across the junction varies linearly with voltage. When two conducting surfaces are brought into intimate contact with a nanoscale junction area, a "point contact" is established between the two surfaces. If the size of the junction is smaller than the mean free path (l) of electrons in the material, the resistance of the contact does not follow Ohm's law.[39] Rather, the electron transport is ballistic, i.e. electron scattering is negligible.

In AFM experiments the typical contact radius is of the order of a few nanometers, which is much smaller than the mean free path of electrons in metals or in graphite. For an electron mean free path much larger than the contact radius of the junction ($l \gg a$), the point contact resistance is given by the Sharvin resistance (R_{Sh}):

$$R_{\mathrm{Sh}} = \frac{4\rho \cdot l}{3\pi \cdot a^2}, \tag{3}$$

where ρ is the mean bulk resistivity. Note that at a fixed applied voltage, the current is proportional to the contact area ($\pi \cdot a^2$). Thus, C-AFM can be used to measure the change in the contact area with the load.

A common situation in AFM is to have a poorly conducting material (e.g. an oxide, protein or organic molecule) sandwiched between the conducting surfaces of the tip and the substrate. Thus, the experimental data need to be modeled as a metal–insulator–metal electrical junction. A common approach to modeling the charge transport is to approximate the transmission with a single barrier, as used for dielectric materials. Such approaches are based on Simmon's model for electron tunneling across a barrier,[40] for which the current (I) at low bias voltage (V) is

$$I = \pi a^2 \frac{q^2\sqrt{2m_e\phi}}{h^2 s} e^{-B\cdot s\sqrt{\phi}} V, \tag{4}$$

where q is the electronic charge, h the Planck constant, m_e the effective electron mass, ϕ the barrier height, s the barrier width and B a constant. Importantly, note that again the current is proportional to the contact area at fixed bias and this allows the *relative* change in the contact area ($\pi \cdot a^2$) to be monitored as a function of the load.

For systems in which the intervening medium is a molecular layer, Eq. (4) is usually rewritten as $I = V/R$, with the resistance (R) defined as

$$R = R_0 e^{\beta\cdot s}, \tag{5}$$

where $\beta = B\sqrt{\phi}$ is the decay coefficient for tunneling through the molecule and R_0 is the effective resistance of the molecule–metal contacts. For length scales smaller than the mean free path of electrons, quantum coupling between the orbitals of the molecules occurs at both the metal electrodes and through the molecule itself. The resistance of such a contact junction is described by the Landauer formula:[41]

$$R = \frac{h}{2q^2 NT}, \tag{6}$$

where T is the transmission coefficient, which is determined by the net orbital overlap between electrodes, and N is the number of molecules within the junction (treating the molecules as parallel resistors with only weak cooperative effects). Note that $2q^2/h$ is the quantum unit of conductance. Following a simplified model of Engelkes *et al.* the transmission can be separated into components as $T = T_{\text{tip}} T_{\text{sub}} T_{\text{mol}}$,[42] comprising the tip contact (T_{tip}), the substrate contact (T_{sub}) and the molecule (T_{mol}). Hence, Eqs. (5) and (6) can be combined in the low bias limit as

$$R_0 = \frac{h}{2q^2 N} \frac{1}{T_{\text{sub}}} \frac{1}{T_{\text{tip}}}, \tag{7}$$

$$e^{\beta\cdot s} = \frac{1}{T_{\text{mol}}}. \tag{8}$$

Unfortunately, the transmission coefficients are not sufficiently well known at present to allow a definitive evaluation to be made of the absolute contact area (or, equivalently, N).

6.4. Boundary Lubrication Studies Using C-AFM

Figures 2(a) and 2(b) show raw force–current data taken for hexadecane on graphite using a gold-coated tip. Jumps (labeled $n = 1$–5) are observed as the tip approaches the substrate, corresponding to solvation layering of the confined hexadecane. The tip is certainly in contact with the graphite in the region $n = 0$ as directly observed by AFM imaging, i.e. the lamellar structure of the hexadecane monolayer ($n = 1$) can be imaged at low force whereas the graphite lattice ($n = 0$) appears when the imaging force becomes sufficiently high [Fig. 2(c)]. The simultaneously measured current is shown for the first hexadecane layer [Fig. 2(a)] and within the second layer [Fig. 2(b)]. The raw force curves can be modified to indicate the true tip–substrate separation by subtracting the distance the cantilever is deflected from the piezoelectric displacement [see inset, Fig. 2(a)]. For the second layer the measured current was extremely small and a lock-in amplifier was used to extract the signal from the noise. No current could be observed within the third layer, because the tunnel current was negligible.

In all experiments there is a sudden, discontinuous jump in the current as the tip transverses a solvation layer. A most interesting finding is the presence of two distinct regions within the first layer [Fig. 2(a)], with a sharp increase in current (at -22 nm), just before the tip punches through the hexadecane monolayer to the underlying graphite substrate. The current variation is most usefully plotted as a function of force. We begin by first considering the tip–graphite contact [Fig. 3(a)]. A current is measured over the entire force curve of Fig. 3(a), including the adhesive minima. This is an important and necessary observation, because wear or contamination of the metal coating at the tip apex is a serious problem in C-AFM.[33] A previous study has reported forces of order ~1000 nN to break through supposed "solvation layers" of hexadecane on graphite using C-AFM.[43] Given our data, it is not realistic that the multiple current "jumps" observed

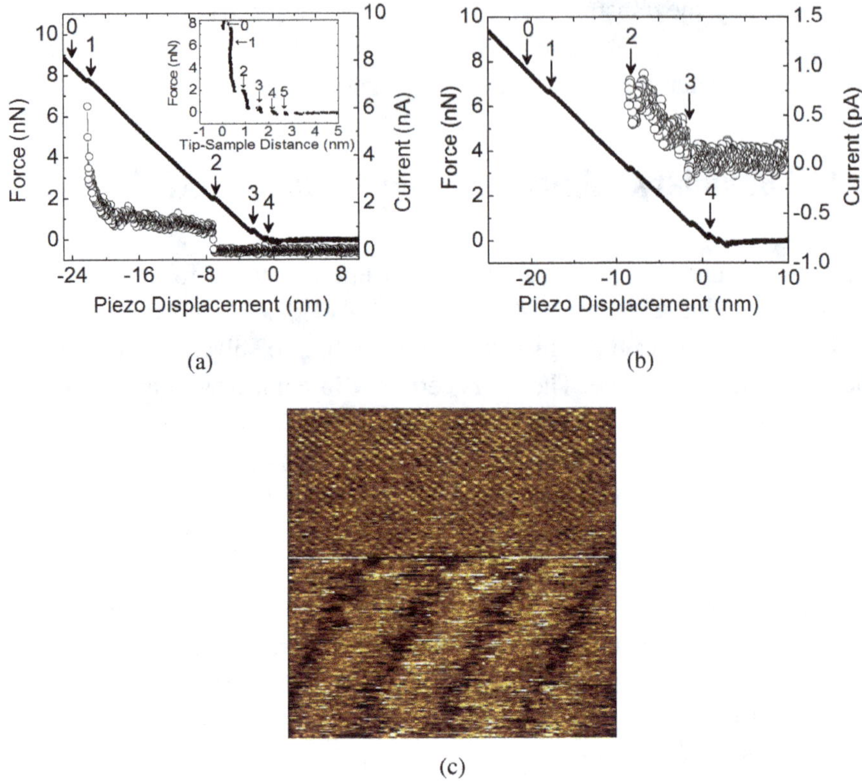

Fig. 2. Simultaneous force (solid line) and current (circle) measurements for hexadecane on graphite as a function of displacement of the piezoelectric actuator. Current is shown for the (a) $n = 1$ layer and (b) $n = 2$ layer. The tip is approaching the surface. Solvation "jumps" are observed in both the force and current curves and are labeled $n = 0$–4, with $n = 0$ being the tip in contact with the graphite substrate. The inset shows the force as a function of the tip–sample separation. (c) 12.5 nm×12.5 nm contact mode AFM topographic image of hexadecane adsorbed on graphite (bottom half). Increasing the force allows the graphite lattice ($n = 0$) to be imaged (top half).

in Ref. 43 are associated with solvation layering, and we believe that the underlying reason for this misinterpretation is wear or contamination of the tip apex.

We now model the tip–graphite contact. In all cases we consider, the measured current is proportional to the area. Thus, a plot of current versus force shows the *relative* change in the area with the applied force, which

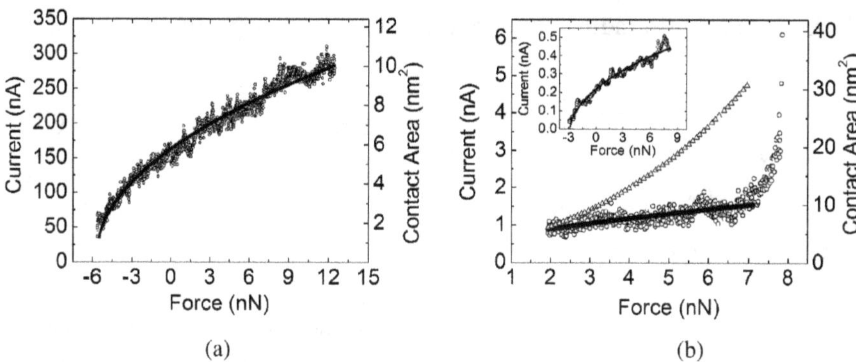

Fig. 3. (a) Current–force curve for the tip in contact with the graphite ($n = 0$). The tip is being pulled off the surface. The variation in current is fitted with the Maugis–Dugdale model (solid curve) to give the contact area. (b) Current–force curve for the tip in contact with the first hexadecane layer ($n = 1$). There are distinct "slow current" and "fast current" regions. A DMT profile is superimposed (solid curve) to estimate the mechanical contact area. The inset shows data taken with a different tip, as the tip is pulled off the first layer. The variation of current follows DMT mechanics with a very small contact area ($\sim$0.7 nm^2) at the adhesive minima ($-$2.9 nN). The curve (Δ) shows the expected current variation if the confined molecules are assumed to undergo significant deformation upon compression [i.e. as in Eq. (11)], which is clearly not the case.

in turn can be compared with the contact area calculated from the applied force coupled with a suitable mechanical model of the point contact. We apply the MD model [Eq. (1)] and find a good fit to the force–current data [solid line, Fig. 3(a)] with a value of $\alpha = 0.073$, i.e. close to the DMT model. Thus, the *relative* change in the contact area is well described for the hexadecane–graphite junction by elastic continuum mechanics.

The *absolute* value of the contact area can be estimated from either fitting mechanical models, as shown in Fig. 3(a), or the current measurement. By the use of Eq. (3), the calculated value of the "electrical" contact area for the data of Fig. 3(a) at zero applied force is $\sim$5.8 nm^2, using $\rho \sim 5000\,\mu\Omega$m[44] and $l = 10$ nm. We calculate the mechanical area using the DMT model ($\alpha = 0$), which gives values only slightly larger than those found using the full MD model. The DMT contact radius is

$$a_{\mathrm{DMT}} = \left\{ \frac{R_{\mathrm{tip}}}{K}(F + F_c) \right\}^{1/3}, \tag{9}$$

where R_{tip} is the tip radius of curvature and K is the effective modulus,

$$K^{-1} = \frac{3}{4}\left(\frac{1-\nu_1^2}{E_1} + \frac{1-\nu_2^2}{E_2}\right), \tag{10}$$

with E being Young's modulus, ν the Poisson ratio, and the subscripts 1, 2 referring to gold and graphite respectively. Using $\nu_1 = 0.33$, $\nu_2 = 0.43$, $E_1 = 78\,\text{GPa}$ and $E_2 = 36.5\,\text{GPa}$ (for graphite along the c axis) results in $K = 39.5\,\text{GPa}$. For the data of Fig. 3 we measure $R = 25\,\text{nm}$ (using scanning electron microscopy), giving the calculated "mechanical" contact area at zero applied force as $7.4\,\text{nm}^2$.

The electrical and mechanical estimates of the contact area ($5.8\,\text{nm}^2$ and $7.4\,\text{nm}^2$ respectively) are close. This is probably fortuitous, because the uncertainties are large. All that can be stated is that *qualitatively* the electrical and mechanical approaches give the same order of magnitude in contact area. A consideration of the errors involved leads to the conclusion that the mechanical model is best, because the largest error arises from the mean free path in the "electrical" calculation, which can vary by an order of magnitude.

Having verified that the elastic model is reasonable for $n = 0$, we now consider the first hexadecane layer ($n = 1$). The force–current curve [Fig. 3(b)] shows two distinct regions, which we label the "fast" current region, designating the very rapid change in current at high force, and the "slow" current region, occurring at lower forces. The current flow across the hexadecane layer occurs by tunneling[45] at low bias, following Eq. (5). By fitting the distance decay of the resistance values found for the $n = 1$ and $n = 2$ layers, we find that $\beta \approx 1.8\,\text{Å}^{-1}$ and $R_0 \approx 180\,\text{k}\Omega$.

The current at constant voltage is proportional to the contact area, allowing *relative* changes to be monitored. However, for the first layer, the *absolute* value of the "electrical" contact area cannot be accurately evaluated at present,[42] because the electron transmission function is not adequately known for alkanes, and the mechanical modeling again proves the most reliable for evaluating the absolute contact area. In the slow current regime we find that the DMT model provides a good description of the mechanical response of the monolayer [see inset, Fig. 3(b)], suggesting that the small increase in current arises chiefly from an increase in the contact area, i.e. we observe that $I \propto a^2 \propto (F + F_c)^{2/3}$, as expected from Eq. (9). The calculated contact area varies from ~5.8 nm^2 at 2 nN to ~10.2 nm^2 at 7 nN, using an effective modulus of $K = 39.5$ GPa.

The modulus of the tip–graphite contact (39.5 GPa) is used because the deformation of the hexadecane monolayer is negligible in comparison with the deformation of the tip or graphite, as we now show. If we assume that the hexadecane molecules are more compliant than either gold or graphite, the current variation will be dependent on the deformation in the confined monolayer in addition to changes in the contact area. Deformation of the monolayer leads to an exponential increase in current, because the tip-to-substrate distance will decrease with an increasing load. The resulting variation in tunnel current as a function of force for the DMT model would be[46]

$$I \propto (F + F_c)^{2/3*} \exp((F + F_c)^{2/3}). \quad (11)$$

This current variation is shown qualitatively as a curve (Δ) in Fig. 3(b) and clearly does not describe the data. We conclude that the hexadecane monolayer is effectively rigid and any monolayer deformation is very small such that there is no significant change in the tip-to-graphite tunnel gap. We believe that the principal deformation under load is occurring in the graphite, and a calculation using the DMT model shows the deformation of the entire tip–monolayer–graphite system to be ~0.13 nm at the maximum force of 7 nN. Again, we note that if the hexadecane monolayer were deforming by 0.13 nm we would expect to observe a ~10-fold increase in the tunnel current, which is not observed.

In the "fast" current region the current rises by a factor of ~4. The variation of current with load cannot be explained by the fitting of a continuum mechanics model. Thus, we rule out the possibility that the contact area suddenly increases by elastic deformation of the substrate or tip. An alternative explanation is that a hole nucleates in the monolayer, prior to the complete squeeze-out.[47] The tip would penetrate into the hole, giving an increased tunnel current. This situation appears unlikely, as molecules squeeze out extremely rapidly once nucleation occurs.[19] Nucleation would describe the rapid *transition* between layers, but not the fast current region of Fig. 3(b).

Our favored model is that the molecules remain under the tip apex in the fast current region, and prior to the complete $n = 1 \rightarrow 0$ layer transition there is a rearrangement of molecules under the tip apex leading to a smaller tunnel gap. This hypothesis ties in with the observation in AFM that the solvation "jump" closest to the surface ($n = 1 \rightarrow 0$) is a smaller displacement than other jumps.[48] In our experiments, outer "jump" distances ($n \geq 2$) are all 4.5 ± 0.4 Å, whereas the $n = 1 \rightarrow 0$ transition is 3.5 ± 0.5 Å, a difference corresponding closely to the expected change in tip–substrate separation of 0.8 Å required for the fourfold increase in current. The conjecture is supported by molecular-dynamic simulations of liquid Xe confined between two solid walls,[49] which show that Xe can undergo a pressure-induced phase transition, leading to a smaller wall–wall separation, prior to a layer transition. A similar two-stage squeezing for the second layer has been noted in simulations of isobutane (though not n-butane) squeezed between two elastic walls.[50]

The conformation change of the confined hexadecane would be expected to occur at the tip apex, because the pressure is higher in the center; for example, for the DMT model the center pressure is a factor of 3/2 higher than the average pressure. For Fig. 3(b) we find that just before the monolayer is removed, the contact area is $\sim$10.7 nm^2, giving the average pressure to squeeze out the layer as $\sim$0.73 GPa, with a pressure at the tip apex of $\sim$1.1 GPa. Not surprisingly, the observed squeeze-out pressure changes due to experimental variation (e.g. approach speed, experimental error, or differing molecular arrangements at the contact) and the use of different tips. The squeeze-out pressure has a statistical distribution, and the median value for the $n = 1 \rightarrow 0$ transition at a 10 nm/s approach speed, with data taken from various tips and repeated experiments, is 0.79 GPa.

An area of 10.7 nm^2 corresponds to $\sim$10 hexadecane molecules in the contact zone. This is much less than for comparable squeezing experiments using the SFA involving contact areas of $\sim$1000 μm^2. In SFA measurements there is a gradual thinning of the confined film as it is squeezed between two surfaces,[51] with regions of n and $n - 1$ layers coexisting within the time frame of the experiment. Such behavior is not expected over the length scale of AFM measurements. The AFM experiments resemble a single asperity contact and are closer to descriptions provided by computer simulations.[47] Indeed, molecular-dynamic computations show that the average squeeze-out pressure for tetradecane confined between two gold surfaces is 0.8 GPa for the $n = 1 \rightarrow 0$ transition,[52] which is comparable with our experimental data.

6.5. Squeeze-out of Confined Branched Molecules

Now, we extend the study to branched molecules. Figure 4(a) shows representative force curves taken in squalane and 2,2,4,4,6,8,8-heptamethylnonane (HMN). For squalane, ~5 solvation jumps are clearly observed, as indicated by the labels $n = 1$–5, where $n = 0$ corresponds to the tip in direct contact with the graphite substrate and $n = 1$ is the monolayer. In HMN, we observe weaker solvation layering (labeled $n = 1$–3) and most of the force curves do not reveal sharp jumps between layers. The transition between solvation layers resembles that of yielding of a polymer monolayer,[53] with small kinks (shown by unlabeled arrows) observed in the layer closest to the graphite surface ($n = 1$), suggesting a change in conformation of the confined molecules under compression.

We first consider the squalane data. Several current–force curves obtained at 25°C using the same tip on the first monolayer ($n = 1$) are shown in Fig. 4(b). All the curves show similar variation with load, as observed for a hexadecane monolayer on graphite [Fig. 3(b)]. We can model the data as explained for the hexadecane–graphite system. This indicates that squalane also behaves as an elastic, solid-like monolayer at 25°C. We confirm this by STM imaging of the squalane monolayer, which reveals ordered domains of lamellar structures [Fig. 4(c)]. The molecular resolution images reveal individual squalane molecules aligned parallel to each other. The lamellar spacing is ~4.0 nm and we observe a diffuse boundary between the squalane lamellar stripes. This contrasts with linear alkanes such as hexadecane and tetracosane (a molecule with a chain length similar to that of squalane), where very sharp lamellar boundaries are observed.[54] These images of squalane support recent simulations of such surface ordering,[29] in which the authors have also shown that previous experimental studies based on diffraction methods[26,27] are unable to confirm ordering in squalane due to broadening of the diffraction peaks.

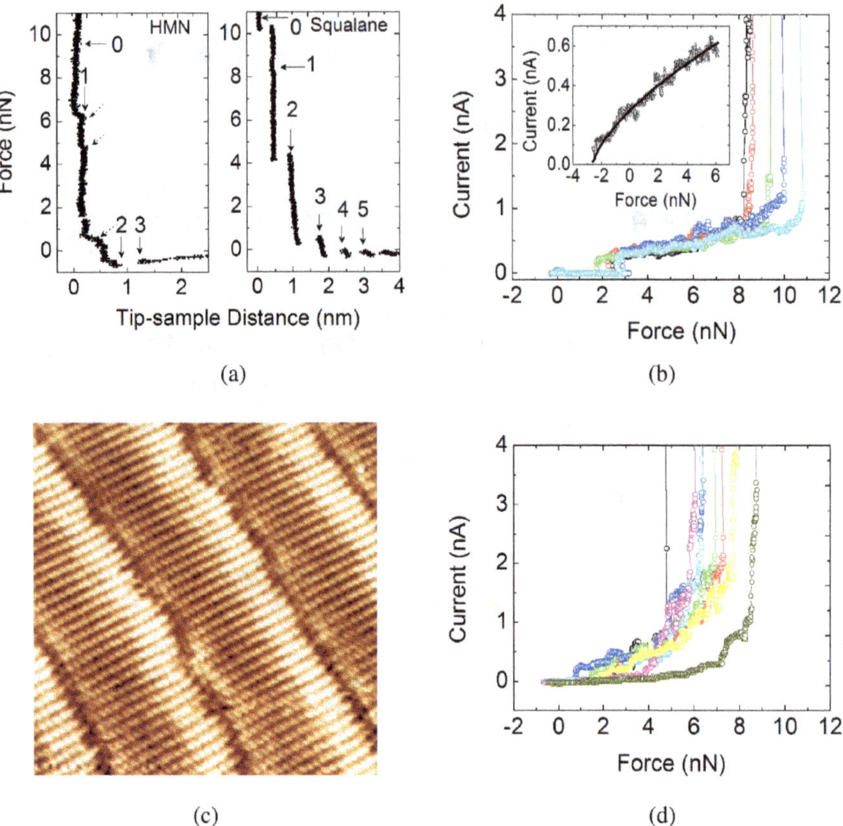

Fig. 4. (a) Data showing force as a function of the tip–sample separation for squalane and HMN on graphite. Clear solvation jumps are observed in squalane, indicated by $n = 0 - 5$ ($n = 0$ is the graphite surface). HMN shows less defined jumps, indicated by $n = 0 - 3$ ($n = 0$ is the graphite surface), with several kinks in the force curve (shown with unlabeled arrows). (b) Current–force curve for the tip in contact with the first squalane layer ($n = 1$). There are distinct "slow-varying current" and "fast-varying current" regions (the latter being close to the $n = 1 \rightarrow 0$ layer transition at 8–10 nN force). The inset shows data taken with the same tip as the tip is pulled off the first layer, indicating that the variation of current with force follows a DMT model. (c) Molecular resolution STM topographic image revealing individual squalane molecules. Image size = 13 nm × 13 nm. Tunnel conditions: $V = 1.0$ V (sample positive); $I_t = 12$ pA. (d) Current–force curve for HMN at low force. The tip is within the monolayer ($n = 1$) over most of this force region (there is uncertainty at the higher forces about whether the tip is within $n = 1$ or $n = 0$).

We now compare force curve data of HMN with squalane and hexadecane to reveal fundamental differences between these confined fluids. Figure 4(d) shows current–force measurements at 25°C where the tip is probing the HMN monolayer ($n = 1$). Compared to squalane, the current signals do not show a systematic variation with force; the squeeze-out force for the $n = 1 \rightarrow 0$ transition varies considerably (by a factor of 2); and we find that the data cannot be fitted to any generalized model of an elastic point contact, such as the MD model.

Figure 5 shows current–force data for squalane and HMN with the tip in contact with the graphite substrate ($n = 0$). In squalane at 25°C [Fig. 5(c)] the current–force curves are reversible in loading and unloading cycles. Assuming that the measured current is proportional to the contact area, we find that the DMT model [Eq. (9)] again provides a good fit to the data. Thus, as shown for contacts in hexadecane (Fig. 3), the relative change in the contact area of the tip–graphite junction is well described by an elastic continuum model. The tip–graphite contact mechanics is very

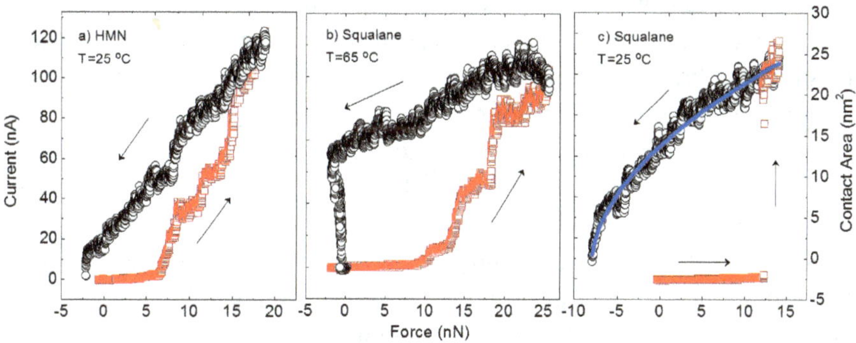

Fig. 5. Current–force curves at high force. The arrows indicate the direction of the force curve cycle. The tip–sample approach speed is 10 nm/s. (a) Data for HMN at 25°C. The curve does not show a sharp squeeze-out of the monolayer ($n = 1 \rightarrow 0$ transition), although consideration of the current magnitude indicates that some part of the tip apex is in contact with graphite at forces $\geq$10 nN. (b) Data for squalane at 65°C. The curve does not show a sharp squeeze-out of the monolayer ($n = 1 \rightarrow 0$ transition) and cannot be fitted to a continuum elastic model. (c) Data for squalane at 25°C. The monolayer is squeezed out suddenly ($n = 1 \rightarrow 0$ transition) and the tip contacts the graphite when the force reaches ~12.5 nN (squares). The variation in current while pulling the tip off the graphite surface (circles) is fitted with the DMT model (curve) to give the contact area.

different in HMN [Fig. 5(a)]. We observe a gradual variation of current, even up to forces of ∼20 nN, where the junction resistance in HMN (137 ± 65 MΩ per nm^2) approaches the resistance found in squalane (77 ± 21 MΩ per nm^2) — an observation which gives confidence that some part of the tip is indeed in contact with the graphite surface. The current variation with load cannot be fitted to elastic contact models and does not follow a reversible path. Significant hysteresis and variability is observed between the approach and retraction cycles. It is significant that we cannot fit elastic continuum models to the current–force data for the tip in contact with the graphite [Fig. 5(a)]. This result is surprising and implies that in this case the state of the material near or within the junction negates the use of point contact models developed for simple elastic solids.

We believe that the trends in the HMN data (hysteresis; nonuniform current variation with force) arise because the HMN molecules are in a liquid-like state. An explanation of the differences in the observed force curves [for example, between Fig. 5(a) and Fig. 5(c)] can be found in recent simulations comparing the squeeze-out of a linear alkane (butane) and its branched isomer, isobutane.[50] The linear molecules form an ordered monolayer and are completely removed from the contact zone under applied pressure. The branched isomer (isobutane) remains liquid-like in the contact zone under identical conditions of temperature and pressure and shows a higher resistance to displacement, leading to the trapping of a few molecules, even at very high pressure.[50] Essentially, the confined molecules display viscoelastic behavior and are displaced only slowly from the gap. If the pressure becomes sufficiently high before the molecules can be displaced, then the confining surfaces will deform, enabling hollows filled with isobutane to be created. We confirm this interpretation by drifting the tip extremely slowly (∼1.0 nm/s) toward the surface, waiting ∼10 s at a high applied load, and finally pulling the tip off the surface as in a routine force curve. We observe a force curve which is similar to a tip in contact with graphite in squalane [i.e. a pulloff curve similar to Fig. 5(c)], demonstrating that a solid–solid contact *can* be formed in HMN if the loading rate is very slow. Thus, trapping of HMN molecules under the tip during compression appears a plausible mechanism.

Strong support that the change in the state of the confined material (liquid-like or solid-like) is the underlying cause is given by experiments at different temperatures. Figure 5(b) shows a representative force curve undertaken in squalane at 65°C (note: the entire liquid cell is isothermal), with the tip in mechanical contact with the graphite substrate ($n = 0$). The force curve behavior resembles that of HMN at 25°C [Fig. 5(a)]. Further, despite repeated attempts, no STM image of the squalane monolayer at 65°C could be obtained, strongly suggesting the monolayer had ceased to be ordered. This assertion is supported by a recent simulation which showed that the solid phase of the squalane monolayer on graphite melts at ~52°C.[29] Similarly, we have observed that the hexadecane monolayer on graphite is well ordered at 25°C and shows solid-like squeeze-out (Fig. 3), whereas at 65°C the monolayer cannot be imaged and shows force curves similar to Fig. 5(b). It is known that the hexadecane monolayer on graphite melts at ~55°C.[55]

Thus, the state of the confined material, not the branching, is the key condition dictating the squeeze-out behavior and dynamics.[56] The degree of branching and the temperature influence the specific state of the monolayer. Note that graphite is not a special substrate for the formation of ordered molecular layers. Linear alkanes also form ordered monolayers on a variety of other substrates having a significant lattice mismatch, e.g. MoS_2, $MoSe_2$ and Au(111),[57,58] indicating that surface flatness is most important and the graphite lattice has small effect on the surface ordering of alkane monolayers.

6.6. Conclusions and Outlook

This chapter has detailed the behavior of simple liquids on graphite. We found that the squeeze-out at 25°C of "solid-like" monolayers of linear alkanes (e.g. hexadecane) and the branched alkane, squalane, can be described by continuum mechanics models for an elastic solid. This is for the tip both in contact with the underlying substrate and within the solvation layers. The use of C-AFM shows that just before the squeeze-out of the solid monolayer there is a characteristic subtle rearrangement of the molecules under the tip apex, in agreement with computer simulations.[49] The solid-like nature of the hexadecane or squalane monolayer on graphite at 25°C can be verified by direct imaging using STM or AFM.

In contrast, 2,2,4,4,6,8,8-heptamethylnonane (HMN), a more densely branched alkane, remains disordered on the surface. Molecular scale STM images of this disordered system could never be obtained. Although both squalane and HMN are strongly branched alkanes, the disorder/order distinction in the state of the confined monolayer leads to striking differences in the solvation layering and squeeze-out behavior of the two molecules at room temperature. Squeezing of HMN reveals significant variability in data due to its disordered state. Surprisingly, continuum elastic models cannot be applied to describe the contact, either on the monolayer or with the tip in contact with the graphite. We postulate that "liquid-like" HMN molecules always remain trapped within the tip–sample junction, even at very high applied loads, as suggested in a recent simulation showing trapping of molecules under nanoscale confinement.[50] The mechanism of trapping of the disordered molecules within the nanoscopic contact was confirmed by repeating the current versus force measurements at different speeds and at elevated temperatures (above the monolayer melting temperature), which changed the hexadecane and squalane data from solid-like behavior (at low temperature) to that observed for disordered HMN molecules.

The work presented in this chapter shows that various factors, such as shape of molecules, temperature or speed of experiments, can have significant effect on solvation forces and squeeze-out behavior of confined materials. This opens up numerous possibilities of exploring confined liquids with other AFM techniques and studying different systems of interest to tribology and biology. For example, dynamic mode AFM can be used to give information about the viscosity of the confined liquid.[59] Measuring the effect of molecular shape, temperature and speed of approach on viscosity with dynamic mode AFM will provide critical information on the timescales of the molecular motion and lead to quantification of the terms "solid-like" and "liquid-like." It is also possible to combine conductivity measurements with dynamic mode AFM measurements.

This work has shown that C-AFM can be used to provide insight into the mechanics of single asperity contacts, in addition to the measurement of force curves. Qualitatively, the variation of current with load is well understood. However, more theoretical effort is required to enable strong quantitative predictions to be made, in particular for the evaluation of the absolute tip–sample contact area from electrical measurements.

Experimentally, C-AFM would be more versatile if robust metallic tips were available. At present, only low voltages can be applied and scanning is kept to a minimum to prevent wear. The fabrication of more robust tips would enable, for example, the measurement of current in the outer solvation layers ($n > 2$) by increasing the applied bias voltage. The current flow during scanning could also be measured and thus the current could be measured simultaneously with the friction forces acting on the point contact.

References

1. K. L. Johnson, *Contact Mechanics* (Cambridge University Press, 1985).
2. J. N. Israelachvili, *Surf. Sci. Rep.* **14**, 109 (1992).
3. K. L. Johnson, *Proc. R. Soc. London, Ser. A–Math. Phys. Eng. Sci.* **453**, 163 (1997).
4. H. Hertz and J. Reine, *Angew. Math.* **92**, 156 (1882).
5. B. V. Derjaguin, V. M. Muller and Y. P. Toporov, *J. Colloid Interface Sci.* **53**, 314 (1975).
6. K. L. Johnson, K. Kendall and A. D. Roberts, *Proc. R. Soc. London, Ser. A* **324**, 301 (1971).
7. R. G. Horn and J. N. Israelachvili, *J. Chem. Phys.* **75**, 1400 (1981).
8. G. Binnig, C. F. Quate and C. Gerber, *Phys. Rev. Lett.* **56**, 930 (1986).
9. B. V. Deryaguin and L. D. Landau, *Acta Physico-Chem. U.R.S.S.* **14**, 633 (1941).
10. E. J. W. Verwey and J. T. G. Overbeek, *Theory of the Stability of Lyophobic Colloids (Elsevier*, 1948).
11. J. N. Israelachvili, *Intermolecular and Surface Forces* (Academic, New York, 1992).
12. J. C. Henniker, *Rev. Mod. Phys.* **21**, 322 (1949).
13. S. J. O'Shea, M. E. Welland and T. Rayment, *Appl. Phys. Lett.* **60**, 2356 (1992).
14. S. J. O'Shea, M. E. Welland and T. Rayment, *Appl. Phys. Lett.* **61**, 2240 (1992).
15. A. Maali *et al.*, *Phys. Rev. Lett.* **96**, 086105 (2006).
16. S. J. O'Shea, M. E. Welland and J. B. Pethica, *Chem. Phys. Lett.* **223**, 336 (1994).
17. S. Patil *et al.*, *Langmuir* **22**, 6485 (2006).
18. H. J. Butt, B. Cappella and M. Kappl, *Sur. Sci. Rep.* **59**, 1 (2005).
19. H. J. Butt and V. Franz, *Phys. Rev. E* **66**, 031601 (2002).
20. T. Fukuma, M. J. Higgins and S. P. Jarvis, *Biophys. J.* **92**, 3603 (2007).
21. R. Lim and S. J. O'Shea, *Phys. Rev. Lett.* **88**, 246101 (2002).
22. J. N. Israelachvili *et al.*, *Macromolecules* **22**, 4247 (1989).
23. Y. X. Zhu and S. Granick, *Phys. Rev. Lett.* **93**, 096101 (2004).
24. D. Gourdon and J. Israelachvili, *Phys. Rev. Lett.* **96**, 099601 (2006).
25. J. S. Wong *et al.*, *Phys. Rev. Lett.* **96**, 099602 (2006).

26. D. Fuhrmann and A. P. Graham, *J. Chem. Phys.* **120**, 2439 (2004).
27. D. Fuhrmann *et al.*, *Surf. Sci.* **482**, 77 (2001).
28. H. D. Mo, G. Evmenenko and P. Dutta, *Chem. Phys. Lett.* **415**, 106 (2005).
29. A. D. Enevoldsen *et al.*, *J. Chem. Phys.* **126**, 104703 (2007).
30. M. A. Lantz *et al.*, *Phys. Rev. B* **55**, 10776 (1997).
31. R. W. Carpick, D. F. Ogletree and M. Salmeron, *Appl. Phys. Lett.* **70**, 1548 (1997).
32. M. A. Lantz *et al.*, *Appl. Phys. Lett.* **70**, 970 (1997).
33. M. A. Lantz, S. J. O'Shea and M. E. Welland, *Phys. Rev. B* **56**, 15345 (1997).
34. I. Szlufarska, M. Chandross and R. W. Carpick, *J. Phys. D.* **41**, 39 (2008).
35. B. Q. Luan and M. O. Robbins, *Nature* **435**, 929 (2005).
36. N. N. Gosvami *et al.*, *J. Chem. Phys.* **126**, 214708 (2007).
37. D. Maugis, *J. Colloid Interface Sci.* **150**, 243 (1992).
38. R. W. Carpick, D. F. Ogletree and M. Salmeron, *J. Colloid Interface Sci.* **211**, 395 (1999).
39. Y. Sharvin, *Sov. Phys. JETP* **21**, 655 (1965).
40. J. G. Simmons, *J. Appl. Phys.* **35**, 2655 (1964).
41. R. Landauer, *Phys. Lett. A* **85**, 91 (1981).
42. V. B. Engelkes, J. M. Beebe and C. D. Frisbie, *J. Am. Chem. Soc.* **126**, 14287 (2004).
43. D. L. Klein and P. L. McEuen, *Appl. Phys. Lett.* **66**, 2478 (1995).
44. G. Kaye and T. Laby, *Tables of Physical and Chemical Constants* (Longman, London, 1973).
45. D. M. Cyr, B. Venkataraman and G. W. Flynn, *Chem. Mater.* **8**, 1600 (1996).
46. X. D. Cui *et al.*, *Nanotechnology* **13**, 5 (2002).
47. B. N. J. Persson and F. Mugele, *J. Phys. Condens. Matter* **16**, R295 (2004).
48. H. H. Butt and R. Stark, *Colloids Surfaces A–Physicochem. Eng. Aspects* **252**, 165 (2005).
49. B. N. J. Persson and P. Ballone, *J. Chem. Phys.* **112**, 9524 (2000).
50. U. Tartaglino *et al.*, *J. Chem. Phys.* **125**, 014704 (2006).
51. T. Becker and F. Mugele, *Phys. Rev. Lett.* **91**, 166104 (2003).
52. I. M. Sivebaek, V. N. Samoilov and B. N. J. Persson, *J. Chem. Phys.* **119**, 2314 (2003).
53. S. J. O'Shea, M. E. Welland and T. Rayment, *Langmuir* **9**, 1826 (1993).
54. L. Askadskaya and J. P. Rabe, *Phys. Rev. Lett.* **69**, 1395 (1992).
55. P. Espeau and J. W. White, *J. Chem. Soc.–Faraday Trans.* **93**, 3197 (1997).
56. N. N. Gosvami, S. K. Sinha and S. J. O'Shea, *Phys. Rev. Lett.* **100**, 076101 (2008).
57. S. Cincotti and J. P. Rabe, *Appl. Phys. Lett.* **62**, 3531 (1993).
58. H. M. Zhang *et al.*, *Chemistry–A Eur. J.* **10**, 1415 (2004).
59. M. J. Martin, H. K. Fathy and B. H. Houston, *J. Appl. Phys.* **104**, 8 (2008).

Nitya Nand Gosvami obtained a BTech with honors in Metallurgical Engineering at the Institute of Technology, Banaras Hindu University, India in 2003, and a PhD in Mechanical Engineering at the National University of Singapore in 2008. He joined the Leibniz Institute for New Materials in Germany as a Postdoctoral Scientist in March 2008. Currently he is a recipient of the Alexander von Humboldt fellowship. During April–May 2008, he was a visiting scientist in the Department of Physics at McGill University in Canada. His research interests include friction, elastic and plastic deformation of materials, and charge transport behavior at the nanometer and molecular scales using atomic force microscopy.

Sean O'Shea received his BSc and PhD in Physics from Sydney University, Australia. Subsequently he undertook research in various areas of nanoscale science at Cambridge University, UK from 1989 to 1998, where he was appointed a Royal Society University Research Fellow. Since 1999 he has been at the Institute of Materials Research and Engineering (IMRE), Singapore. His research interests are in nanotechnology (chiefly scanning probe microscopy), MEMS, and the creation and application of new biosensors.

Dynamic Force Microscopy in Liquid Media

Chapter

7

Wulf Hofbauer*

Abstract

In atomic force microscopy, a probe brought into contact or near-contact with a sample "feels" the surface via mechanical interaction forces. This is appealing, because the experimental data are intuitively understandable. More importantly, this places no special restrictions on the nonmechanical properties of the sample or its environment, making it an extremely versatile method.

Compared to quasi-static force microscopy schemes, dynamic techniques offer enhanced sensitivity and resolution. In ultrahigh vacuum, atomic resolution is almost routine.[1,2] However, a dense liquid environment greatly affects the dynamic properties of the AFM probe. Dynamic AFM techniques in liquid media therefore pose unique challenges, and molecular/ atomic resolution has been achieved only recently.[3,4]

This chapter discusses some of the instrumentation challenges encountered in the conversion of a commercial AFM instrument for high-resolution, frequency-modulated AFM (FM-AFM) in a liquid environment. Both imaging and spectroscopy are demonstrated in application examples. The chapter concludes with a brief outlook on the prospects of dynamic AFM for the study of biomaterials.

*Institute of Materials Research and Engineering, Agency for Science, Technology and Research (A*STAR), 3 Research Link, Singapore 117602.
Email: wulf-h@imre.a-star.edu.sg

7.1. Introduction

In conventional contact and force modulation AFM techniques, the timescale of force measurements is comparatively long, and the mechanical reaction of the AFM probe can be considered as essentially instantaneous. This means that the dynamics of the probe can be ignored, and only quasi-static equilibrium forces are measured. On a much faster detection timescale, the dynamic response of the probe, i.e. inertial effects, cannot be neglected. The acceleration of the probe, governed by its intrinsic dynamic response and by the interaction with the sample, becomes the dominant feature.

The effective timescale on which dynamic effects occur is given by the mechanical resonance frequency of the AFM cantilever. Quasi-continuous force measurements — necessary for scanned operation — require rapid repetition of the elementary dynamic measurement. The natural approach is to oscillate the probe at or near its mechanical resonance frequency, and derive the tip–sample interaction forces from the perturbation of the oscillation characteristics upon engaging the sample.

Rather than directly evaluating acceleration as a function of time,[5] it is usually more convenient to describe and analyze the cantilever vibration in the frequency domain. For sufficiently slow scanning of the probe (adiabatic changes of the oscillation parameters), the oscillation approaches the strictly periodic steady-state solution, and traditional, readily available narrow-band techniques (such as lock-in detection) can conveniently be used. In principle, the time-dependent tip–sample interaction force can be completely reconstructed (over the range covered by the oscillation amplitude) from the measured Fourier coefficients (amplitudes and phases) of all harmonics, either by reconstructing the probe acceleration via an inverse Fourier transform or via perturbative approaches.[6]

In actual use, for the sake of simplicity and/or practicality, only a small number of harmonics, or even only a single harmonic (most commonly the fundamental), may be measured. Elimination of harmonics successively reduces the amount of detail acquired. This provides a natural path for balancing the amount of acquired spectroscopic information with the complexity of the experimental setup.

7.2. Instrumentation for Operation in Liquid

To realize the special challenges of dynamic mode operation in liquid, it is instructive to consider a vibrating AFM tip in a vacuum environment first. During most of the oscillation cycle, the AFM tip is away from the surface and does not sense the short-range interaction forces contributing to high spatial resolution (Fig. 1). The effect of the tip–sample interaction averaged over the entire cycle, such as the resonance frequency shift, is therefore rather small[6,7] (typical relative resonance frequency changes for noncontact AFM are on the order of 0.01%). However, these small shifts are easily measurable, given the typically extremely narrow resonance curves encountered in vacuum.

Changing to a liquid environment, but leaving everything else the same, the average frequency shift due to tip–sample interaction would be similarly small. However, the surrounding viscous liquid will dampen the dynamic response of the cantilever.

The sharpness or "quality" of the resonance is commonly described by its quality factor,

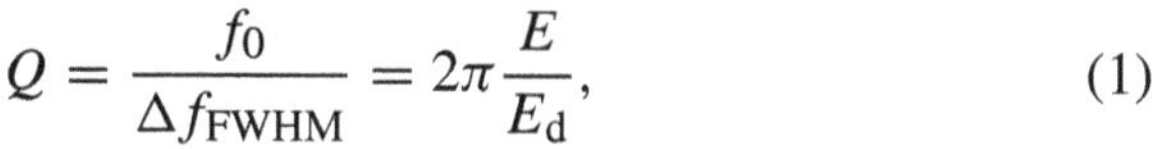

$$Q = \frac{f_0}{\Delta f_{\mathrm{FWHM}}} = 2\pi \frac{E}{E_{\mathrm{d}}}, \qquad (1)$$

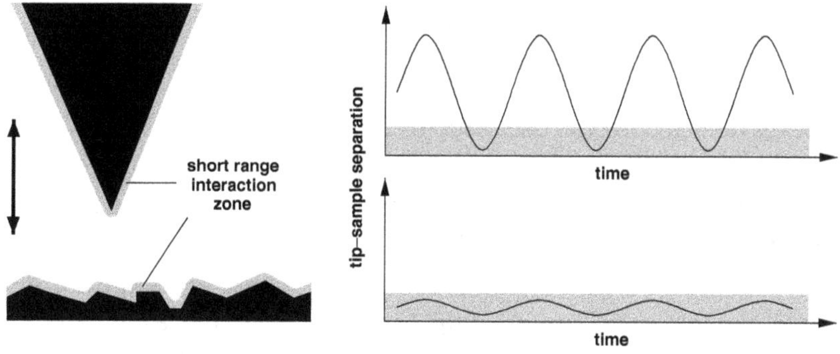

Fig. 1. For large vibration amplitudes, the probe can sense short-range interactions necessary for spatial resolution only during a fraction of the time. Most of the time, the tip interacts with the environment, which may thus have a larger influence on the probe dynamics than the sample. Small amplitude operation overcomes this problem by maximizing the duration of tip–sample interaction.

where f_0 is the resonance frequency, Δf_{FWHM} the -3 dB bandwidth, E the steady-state vibration energy of the cantilever driven at its resonance, and E_{d} the energy dissipation per oscillation cycle. In a liquid environment, E_{d} is strongly increased, so the quality factor Q is vastly reduced, and the resonance becomes very broad.

Measuring frequency shifts that are considerably narrower than the resonance itself requires a very high signal-to-noise ratio (SNR). Furthermore, during an oscillation cycle, the tip is permanently surrounded by the liquid, whereas the interaction time with the sample remains very small. The main contribution to changes in dynamic parameters may therefore not come from the desired tip–sample interaction, but from the tip–liquid interaction instead.

To maximize the influence of the tip–sample interaction, the tip has to be kept within the short-range tip–sample interaction zone. In other words, the oscillation amplitude should be small, ideally in the subnanometer range. This increases the resonance frequency shift and emphasizes the tip–substrate interaction over the tip–liquid interaction. On the other hand, the SNR of the vibration signal deteriorates as amplitudes get smaller. The most critical parameter for sensitive dynamic force microscopy in liquid is therefore the noise level of the cantilever deflection sensor.

7.2.1. *Cantilever readout*

The most common cantilever readout method employs the deflection of a laser beam reflected off the back side of the cantilever (Fig. 2).

The quality of the cantilever deflection sensor is best described by its base noise level, i.e. the cantilever motion equivalent of the electrical noise at the output of the sensor. This also provides an intuitive criterion as to whether a given detector is "good enough": the noise level is ultimately limited by the Brownian motion of the cantilever itself. Once the Brownian motion becomes the dominant noise source in the system, further improvement of the detector yields rapidly diminishing returns. This is also a likely reason why the simple optical beam deflection sensor has not been widely replaced as the standard detection scheme by more refined methods (such

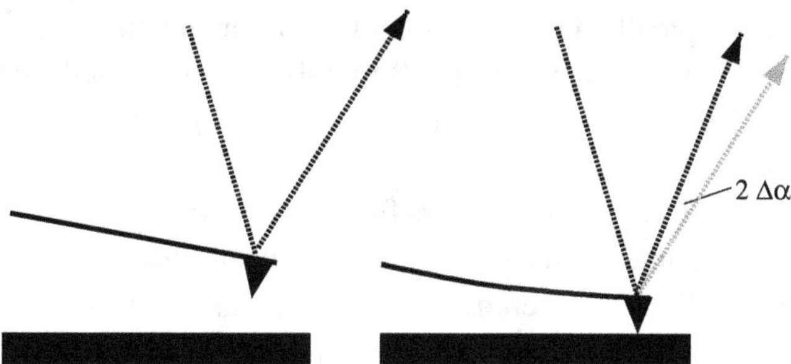

Fig. 2. Scheme of the optical beam deflection sensor. The bending of the cantilever by a small angle $\Delta\alpha$ results in deflection of the sensing laser beam by $2\Delta\alpha$. By projecting the beam on a position-sensitive detector over a distance much larger than the cantilever length, the deflection is geometrically amplified.

as interferometers) — at ambient temperature, there is little to be gained by more sensitive detection schemes.

Nevertheless, it requires some attention to detail in order to exploit the potential of the optical beam deflection sensor. Until recently, most dynamic AFM instruments were intended for operation in vacuum or ambient environment, or comparatively low-resolution operation in liquid, with typical vibration amplitudes of several nanometers. At these large amplitudes, detector noise is fairly uncritical, and there has been little incentive for optimization.

7.2.1.1. *Effects of laser coherence*

Beam-deflection-type detectors register the redistribution of light on a position-sensitive photodetector, caused by geometric reflection off the bending cantilever. Interference patterns also result in energy redistribution, but do not follow the laws of geometrical optics and are therefore an undesirable potential source for noise. Interference effects arise generally from unintended scattering and reflections (multipath interference) within the optical system.

For the instrument modified at the IMRE, interference patterns that change as a result of the cantilever deflection did not appear to contribute significantly to noise. In liquid, the peak–peak cantilever vibration amplitudes are typically a tiny fraction of the laser wavelength, and multi-path interference would manifest itself mainly as a constant dc shift of the cantilever readout, which is generally ignored (and not even detectable by ac-coupled instrumentation).

A much more severe issue arises from coherence-related effects internal to the laser source. Laser resonators can sustain multiple electromagnetic modes at slightly different wavelengths and with differing spatial structures. Depending on the characteristics of the laser medium, several modes may oscillate simultaneously (multimode), or a single oscillation mode may result. Laser operation may spontaneously switch between different modes (mode hopping), with accompanying spontaneous changes of the beam geometry.

In experimenting with various low-cost DVD-player-class laser diodes from different suppliers (wavelength $\lambda \approx 660\,\text{nm}$; optical powers $P \approx 1$–$5\,\text{mW}$), as well as with the unmodified diode laser source of the unmodified, commercial AFM ($\lambda \approx 680\,\text{nm}$; $P < 1\,\text{mW}$), virtually identical behavior was observed. The diodes generally appear to oscillate in a single mode at a time; however, the active mode may hop with slight changes in the device temperature (or other operational parameters). The associated change of beam geometry is interpreted by the optical sensor as a sudden change of the cantilever deflection. In operating regimes where two or more modes compete, rapid, random mode-hopping occurs. The fluctuation of the laser beam structure is registered on the position-sensitive photodiode as vastly increased deflection noise.

Fukuma's seminal article[8] is not clear on whether the mode-hopping noise described is a constant or an intermittent phenomenon. However, with all the laser sources investigated at the IMRE, the mode-hopping noise manifests itself as a temporary (but recurring) effect whenever the laser diode drifts through an unstable operating regime, separated by extended periods of low-noise operation (Fig. 3).

Rapid modulation of the device voltage at frequencies on the order of 300–400 MHz effectively forces the laser diode into multimode oscillation so that the noise-inducing hopping between distinct laser modes is suppressed.[8] When one is projecting the defocused laser beam onto a screen, the accompanying reduced coherence is typically visible to the naked eye as a reduction of the "graininess" of the laser interference speckles, which

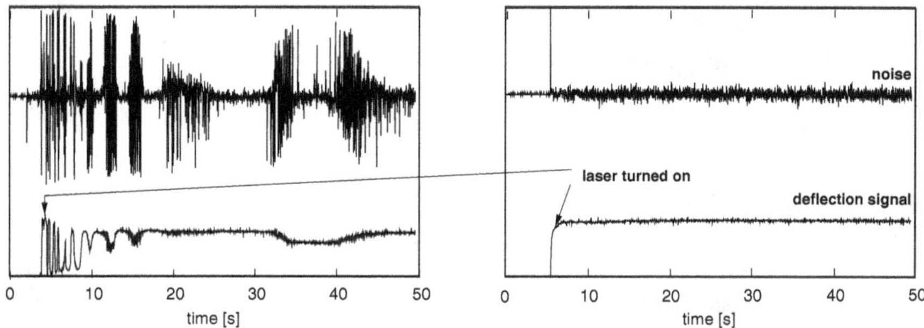

Fig. 3. Demonstration of laser mode-hopping noise. Left: Low-pass filtered deflection signal and noise after switching on a nonmodulated laser diode. The deflection readout changes and the noise increases whenever the laser diode drifts through a regime where several modes compete. Right: When forcing the same laser diode into multimode emission via rf modulation, the signals are stable. The scale is identical in the two graphs.

can be used as an aid to tune the rf modulation of the diode.[a] The reduced speckling should — in principle — be reflected in reduced detector noise, but this was not observed in practice. The principal merit of the rf-induced multimode laser operation lies therefore not in an intrinsic noise reduction compared to a single-mode laser source, but in the circumvention of the unstable mode-hopping regimes.

7.2.1.2. *Effect of the laser numerical aperture*

In beam deflection sensors, the laser beam is focused on the cantilever structure in order to maximize the intensity of the reflected beam. In the case of diffraction-limited optics, large numerical apertures (cone angles of the focused beam) allow better focusing, i.e. smaller laser spot sizes on the cantilever. The downside is that the beam reflected off the cantilever diverges with the same cone angle (assuming that the cantilever back side acts as a flat mirror). The laser spot size on the position detector (typically a quadrant photodiode) ends up rather large in this case. This reduces the sensitivity of the cantilever deflection readout (Fig. 4). Reduction of the numerical aperture, i.e. narrowing of the convergent/divergent cone

[a]While the exact modulation frequency does not matter in principle, the generally impedance-mismatched laser diode results in standing waves on the feed line, making RF power delivery strongly frequency-dependent.

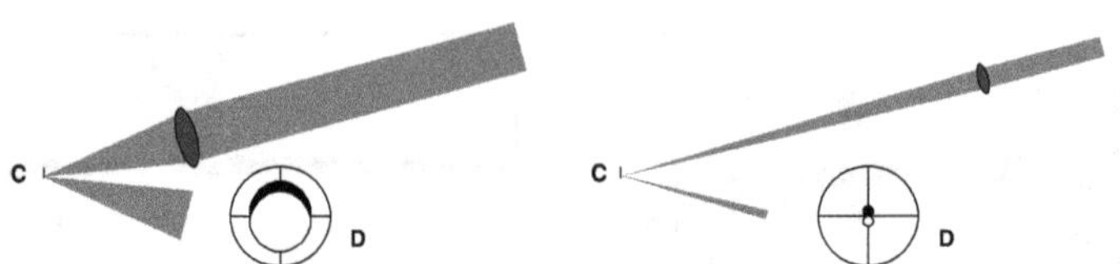

Fig. 4. Left: Large numerical apertures allow better focusing of the laser beam on the narrow cantilever C. However, the rapidly divergent reflected beam leads to a large spot size on the position-sensitive detector D. Only a small fraction of the laser intensity is redistributed for slight bending of the cantilever. Right: By using a low numerical aperture beam, the spot size on the detector is greatly reduced. The same cantilever deflection as on the left results in complete redistribution of the laser energy, greatly enhancing sensitivity.

angle of the laser beam, can therefore achieve markedly improved sensitivity, at the expense of a less well-focused laser spot on the cantilever. Given that typical cantilevers are tens of micrometers wide, the somewhat larger laser beam waist at the cantilever can be tolerated under most circumstances.[b]

Another argument in favor of larger laser spot sizes on the cantilever is that imperfections of the reflecting surface are somewhat averaged out, providing another potential mechanism for noise reduction.

7.2.1.3. *Characterization of noise levels*

A useful description of the deflection sensor noise level is in terms of equivalent deflection noise (random deflections) of a cantilever registered on a (hypothetical) noise-free detector. If the sensitivity S of the detector is defined as the electrical output (in volts) per linear deflection of the cantilever (in nanometers), the rms equivalent noise is simply

$$N_{\text{equiv}} = \frac{N_{\text{elec}}}{S}, \tag{2}$$

where N_{elec} is the electrical noise as measured by an rms voltmeter.

In the case of the optical beam deflection sensor, it should be pointed out that the noise level obtained is a function of both the sensor and the cantilever. This type of sensor really measures the bending angle of the

[b]In informal tests, comparable detector performance was obtained using diffraction-limited aspherical collimation lenses and a focusable collimation lens salvaged from a low-cost laser pointer.

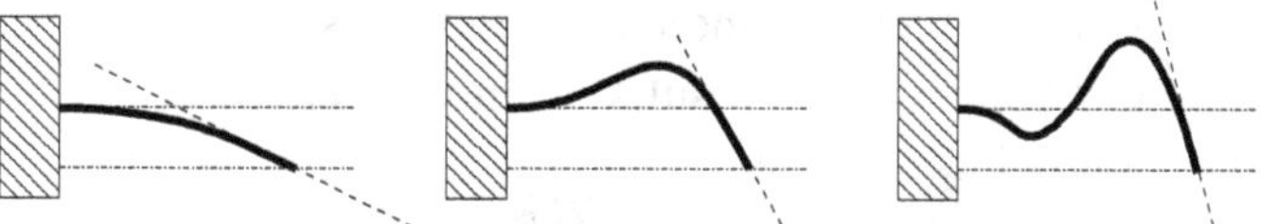

Fig. 5. The first three vibration modes of a cantilever. While the (exaggerated) linear deflection is the same in all cases, the deflection angle differs drastically.

cantilever, not the linear displacement of the tip. Shorter cantilevers will therefore generally result in higher sensitivity S. Furthermore, the relation between displacement, deflection and bending angle may be frequency-dependent: a cantilever vibrating in a higher mode will result in a larger deflection angle (Fig. 5).

The above consideration is valid if the detector noise levels are high. If this is not the case, one has to take into account that part of the observed noise may not be due to the detector, but is caused by the Brownian motion of the cantilever itself. From the equipartition theorem, the average rms deflection N_z of a cantilever with spring constant k_c at temperature T is

$$N_z = \sqrt{\frac{k_\mathrm{B} T}{k_\mathrm{c}}}, \tag{3}$$

where k_B is the Boltzmann constant. If S and k_c are known, the Brownian motion can be accounted for considering that noise amplitudes add geometrically:

$$N_\mathrm{total}^2 = N_\mathrm{detector}^2 + \frac{k_\mathrm{B} T}{k_\mathrm{c}}. \tag{4}$$

While easy to measure, the noise level as described above does not lend itself to comparisons between different detectors, which may have different bandwidths. A better measure is the noise per bandwidth, or spectral rms noise density n (in m/Hz$^{0.5}$). The most convenient method for noise density determination uses a spectrum analyzer. An added advantage of this approach is that noise contributions which are not inherent in the detector, but result from picking up electrical or mechanical interference at specific frequencies, can be easily identified and subsequently remedied.

Modeling the cantilever response around a resonance as a harmonic oscillator, the Brownian motion results in a noise density of[9]

$$n_z(f) = \sqrt{\frac{2k_B T}{\pi k_c Q f_0 \left[1 - (f/f_0)^2\right]^2 + (f/Qf_0)^2}}, \tag{5}$$

where f_0 and Q are the frequency and quality factor of the resonance, respectively. Assuming the detector noise contribution to be frequency-independent (white noise) over the considered spectral range, the measured noise density spectrum is

$$n_{\text{total}}^2 = n_{\text{detector}}^2 + n_z(f)^2, \tag{6}$$

where all relevant parameters can be obtained by fitting to the acquired spectrum (Fig. 6). Conveniently, this also allows to obtain either S or k_c if one of them is not known.

7.2.2. *Cantilever excitation*

The most common way of exciting cantilever vibrations is to incorporate a small piezoelectric slab ("dither piezo") into the AFM cantilever holder assembly. For operation in vacuum and ambient environment, this approach

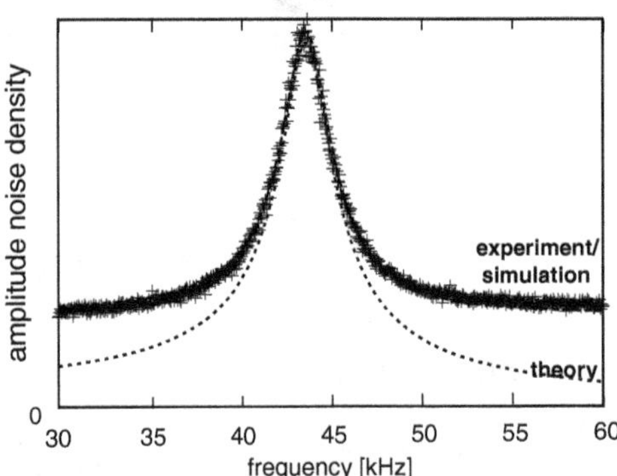

Fig. 6. Brownian motion spectrum of a silicon cantilever ($k_C = 48$ N/m) in water at room temperature. The noise excess compared to the theoretical curve reflects the detector noise (in this case 40 fm/$\sqrt{\text{Hz}}$). Note that it is not necessary to know the scale of the noise density axis, as the necessary information is implicit in the width (quality factor) of the curve.

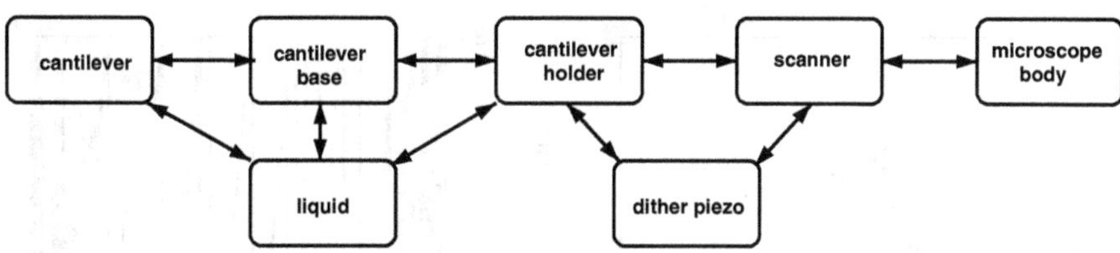

Fig. 7. Flow of acoustic energy in a conventional cantilever excitation scheme using a dither piezo. The microscope parts present themselves as coupled resonators, each with multiple resonance modes of varying frequency and quality factor.

generally works quite well. In liquid media, where the Q factor of the cantilever is greatly reduced, it is problematic.

Figure 7 shows a diagram of how acoustic energy generated by the vibrating piezo slab may travel through a typical microscope setup. Any part of the microscope may form a mechanical resonator, giving rise to one or multiple parasitic resonances.[10,11] Since the resonating parts are very much larger than the cantilever, the acoustic energy stored in these parasitic resonators may be many orders of magnitude larger than the vibration energy of the cantilever itself. These resonances act as filters that modulate the coupling efficiency of the mechanical vibration between the dither piezo and the cantilever. Invariably, the observed response is therefore not that of the cantilever, but a convolution of cantilever and parasitic resonances.

In UHV and ambient environments, the quality factor of the cantilever is typically very high ($Q = 1000$–10,000); in other words, the resonance curve is extremely narrow. There is a very good chance that the narrow resonance will not overlap with similarly sharp parasitic resonances. Parasitics with lower Q will overlap more frequently with the cantilever resonance, but their broader resonance may appear as a more or less straight "baseline" whose influence can be neglected over the narrow interesting frequency range.

The situation in a liquid environment is the reverse: viscous drag on the oscillating cantilever reduces its quality factor to typically $Q \approx 10$. This virtually guarantees spectral overlap between cantilever and parasitic resonances. The much narrower parasitics now become the dominating features of the overall resonance spectrum against the "baseline"

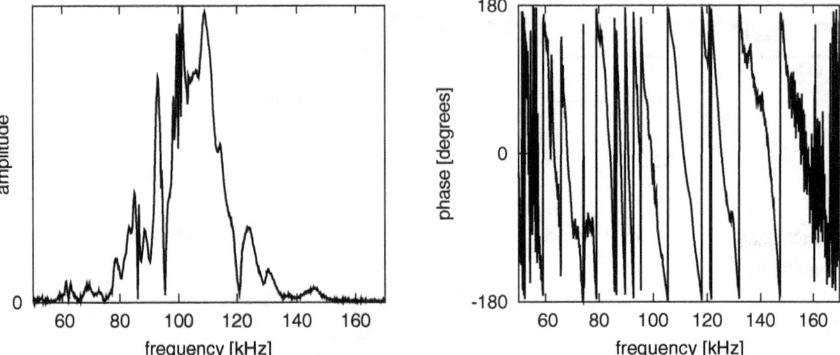

Fig. 8. Resonance curve of a commercial cantilever, excited through the base by a dithering piezo, in water. The amplitude response is heavily marred by parasitic resonances. The phase response — crucial for frequency- and phase-detected AFM — is additionally impaired by a severe slope caused by acoustic delay of the excitation force.

of the cantilever resonance. This effect can be so severe that identifying the cantilever resonance frequency is little more than guesswork. Figure 8 shows a typical resonance curve of a cantilever immersed in water and excited through the base. Note that this case is still relatively benign due to the moderate viscosity of water (as compared to other liquids of interest).

This problem is widespread in commercial setups intended for operation in liquid (such as for biological applications). That it is largely ignored may be due to the overwhelming use of AFM as a nonquantitative imaging technique. If the parasitic resonances do not change during scanning and the cantilever is oscillated at a constant frequency (such as in AM-AFM), their influence may be considered an "instrument constant," and changes in the oscillatory response can be ascribed to the true cantilever dynamics. Even in FM-AFM imaging, the topography feedback loop will adjust the tip height to maintain a constant frequency, and therefore topographic images can be obtained.

In applications where the operating frequency changes (such as frequency shift spectroscopy), the parasitic resonances cannot be ignored: the phase response of the combined cantilever and parasitic resonances is

$$\phi = \phi_c + \phi_p \approx -\pi - \gamma\left[Q_c\left(\frac{f}{f_c} - 1\right) + Q_p\left(\frac{f}{f_p} - 1\right)\right], \tag{7}$$

approximated by a linear expansion around the cantilever/parasitic resonance frequencies $f_{c/p}$. $Q_{c/p}$ are the respective quality factors, and γ is a proportionality constant. If f_c and f are allowed to vary, the phase change is

$$d\phi = \frac{\partial \phi}{\partial f_c} d f_c + \frac{\partial \phi}{\partial f} d f \approx \frac{\gamma Q_c f}{f_c^2} d f_c - \left(\frac{\gamma Q_c}{f_c} + \frac{\gamma Q_p}{f_p} \right) d f. \tag{8}$$

The resonance condition is characterized by a constant phase shift ($d\phi = 0$). From this, and assuming that $f \approx f_c \approx f_p$, one obtains

$$\frac{d f}{d f_c} \approx \frac{1}{1 + \frac{Q_p}{Q_c}}, \tag{9}$$

i.e. the change in overall resonance frequency is attenuated by a factor of $1 + Q_p/Q_c$ relative to the change of the intrinsic cantilever resonance. Unless $Q_p \ll Q_c$, cantilever resonance tracking is systematically wrong, and the apparent frequency shifts are significantly reduced. Since other measurements, such as dissipation, are based on the assumption that the cantilever is maintained at resonance, they are also affected, giving rise to not only quantitative but sometimes even qualitative artefacts. For any application involving variable operation frequencies, it is therefore essential to eliminate the parasitic resonances.

Optimization of the piezo arrangement is of limited potential. To maintain high Q, cantilevers are designed for minimal leakage of acoustic energy to their base. Conversely, transmission of acoustic energy *from* the base *to* the cantilever is equally inefficient. Excitation of the cantilever from the base has to rely mainly on inertial (and, in liquid, viscous drag) forces acting on the cantilever when the base is shaken. The masses (the cantilever base and the supporting clamping mechanism) that need to be vibrated, orders of magnitude larger than the cantilever itself, require correspondingly large forces, whose reaction forces in turn excite the parasitic resonances on the microscope body.

Effective suppression of parasitic resonances therefore requires the exciting force to act directly on the cantilever itself. One possibility is to integrate an actuator into the cantilever, such as by coating with a piezoelectric layer and electrodes.[12] While such specialty cantilevers do exist, the electrical connections, corrosion resistance, etc. pose problems of their own in liquid, potentially conductive, media. Contact-less actuation schemes appear to be clearly more practical. A very elegant method is actuation by photothermal stress,[13–15] which does not require special cantilevers. Electrostatic actuation is another possibility, but is not very practical in conductive media as the electric field is shorted out, and unintentional electrochemical reactions may occur.

An effective alternative (for nonmagnetic samples) is to actuate by magnetostatic forces. A wide selection of commercial cantilevers lend themselves to modification by gluing micrometer-sized magnetic particles to their back.[16,17] After magnetization along the cantilever axis, a (perpendicular) magnetic field B exerts a torque on the magnetic dipole μ of the particle and, indirectly, a vertical force (Fig. 9),

$$F = -\frac{\mathrm{d}(B \cdot \mu)}{\mathrm{d}\alpha} \cdot \frac{\mathrm{d}\alpha(z)}{\mathrm{d}z} = B \cdot \mu \cdot \sin\alpha \cdot \frac{\mathrm{d}\alpha(z)}{\mathrm{d}z}, \tag{10}$$

where $\alpha(z)$ is the angle between the magnetic field and the cantilever end as a function of linear deflection.

While the cantilever modification using micromanipulators under an optical microscope is tedious, the necessary changes to the instrument (inclusion of a small solenoid driving coil) are not difficult, and the resulting cantilever resonance curves fit the simple harmonic oscillator model extremely well (Fig. 9). This not only eliminates the previously discussed

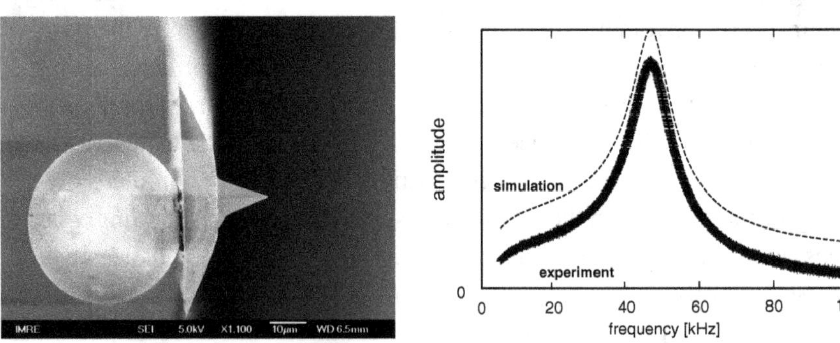

Fig. 9. Left: SEM front view of a silicon cantilever with a spherical magnetic particle attached to the back side. Right: The resonance curve of the magnetically driven cantilever in hexadecane (quality factor $Q = 4.6$) matches the simulation perfectly (simulated curve shifted for clarity). The amplitude falloff at very low frequencies is due to ac coupling of the instrumentation and has been included in the simulation.

artefacts, but also allows one to obtain the resonance parameters with great precision. Another advantage is the frequency independence of the applied magnetic torque. The actuation force can therefore be conveniently calibrated by measuring the static deflection when applying a dc current to the driving coil.

Finally, it should be noted that cantilever actuation through the base leads to complications at low Q factors even in the complete absence of parasitics. The effective excitation force on the cantilever is mediated by inertial and viscous forces and is thus inherently frequency-dependent; furthermore, the vibration described in terms of tip–sample separation is a superposition of the cantilever oscillation relative to its base, and the oscillation of the moving cantilever base relative to the sample/microscope. In the low-Q case, the respective amplitudes may be of similar magnitude, which results in drastic artefacts of the observed vibration amplitude and phase and makes data interpretation challenging.[18]

7.2.3. *Resonance tracking*

The basic premise of FM-AFM is that the cantilever is excited at resonance. When the tip–sample interaction shifts the resonance frequency, the frequency of the excitation signal needs to be corrected accordingly by some kind of feedback scheme.

7.2.3.1. *Self-excitation*

In the simplest case, the amplified and phase-shifted cantilever vibration signal is fed back to the cantilever exciter (Fig. 10).[19] The phase shifter is used to achieve a total shift of 360° over the feedback loop, resulting in self-excitation at the resonance frequency. This scheme is attractive for its simplicity and intuitive dynamic response and is often employed in UHV. In particular, the response to a resonance change is instantaneous, allowing the fastest-possible tracking.

There are some drawbacks to this method. Firstly, noise on the cantilever readout signal will be fed back to the cantilever actuator. Secondly, nonlinear distortions of the detector may be amplified in the direct feedback loop, resulting in harmonics. Thirdly, oscillation is possible using different

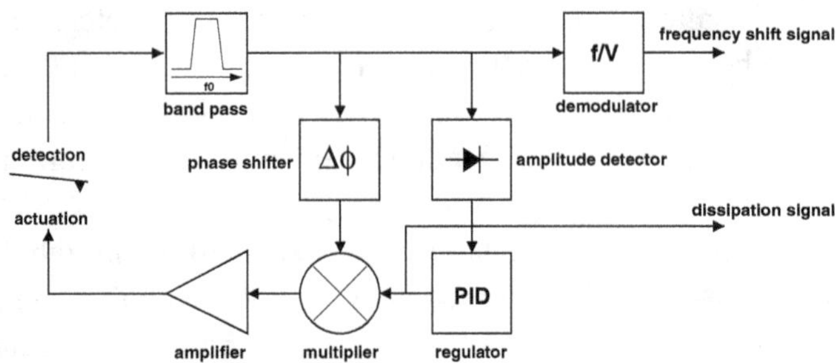

Fig. 10. Self-excitation of the cantilever is the simplest method of tracking resonance frequency changes. Note that both the vibration signal and the noise and distortion of the detector are equally amplified and fed back to the cantilever excitation.

modes of the cantilever, and it is not clear *a priori* that the system will oscillate in the desired mode.

These problems can be addressed by using a bandpass filter centered on the resonance frequency of interest. However, any causal bandpass filter will impart a frequency-dependent phase shift, similar to the parasitic resonances discussed earlier. To minimize this artefact, the bandwidth of the filter must be much larger than the operational frequency shift range, and the filter must have a much lower quality factor than the cantilever [see Eq. (9)]. In small-amplitude operation in liquid, the cantilever quality factor is very low to start with, and the frequency shifts may be rather large. This results in a bandpass filter with such poor selectivity as to be of limited usefulness.

7.2.3.2. *Excitation by a phase-locked loop*

The resonance of a harmonic oscillator is characterized by a phase shift of 90° between exciting force and elongation/deflection. In other words, tracking the cantilever resonance is equivalent to adjusting the driving frequency so as to maintain a constant phase shift. In a phase-locked loop (PLL), the output of a phase detector (which can be analog or digital) is used as feedback signal for correcting the oscillator frequency (Fig. 11).[19]

The PLL approach has several advantages. The signal used to excite the cantilever is completely synthetic and thus free from noise, distortion or other artefacts. At the same time, the low-pass response of the PLL

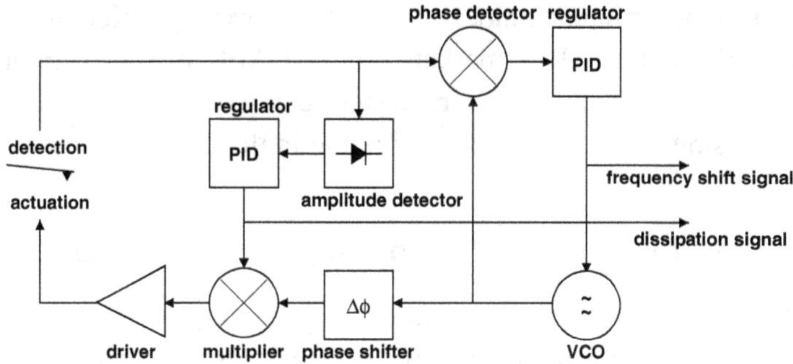

Fig. 11. Cantilever excitation using a phase-locked loop. This scheme intertwines excitation and frequency demodulation. The main advantages are a synthetic, clean excitation signal and the elimination of bandpass filters which interfere with accurate resonance tracking.

internal feedback loop corresponds to a bandpass filter on the cantilever readout signal, reducing noise and suppressing the effects of undesired cantilever modes. However, in contrast to the self-excitation scheme, the center frequency of this equivalent bandpass is given by the PLL operating frequency. As long as the PLL is tracking the cantilever resonance, the bandpass is automatically centered on the resonance frequency of the cantilever. This inherent tracking eliminates the phase shifts associated with normal bandpass filters, and the PLL can accurately follow the cantilever resonance over a wide frequency range while maintaining a much smaller instantaneous bandwidth, thereby reducing noise.

The principal downside of using this scheme is the difficulty of reacquiring a "lock" on the cantilever resonance and re-establishing controlled cantilever oscillation if the PLL has been thrown off-track by some disturbance. When the PLL is part of the topography feedback loop that controls the tip height, momentary loss of the PLL lock can easily result in tip crashes. To counteract this, a function to retract the tip whenever PLL unlocks is virtually a necessity. The detection of the unlocked condition is typically heuristic, based on quantities like measured phase fluctuations, permissible oscillation amplitude limits, etc.

It should be pointed out that most older-generation PLLs have been developed primarily for noncontact AFM in UHV. While the principles of operation are the same, the frequency and amplitude feedback parameters, and criteria for detecting an unlocked condition, are very different for operation in liquid. Not every PLL marketed for FM-AFM use can support such a wide range of adjustments. Modern designs using software-defined digital signal processing are generally more flexible in this respect.

7.2.4. *Frequency modulation vs. phase modulation*

In FM-AFM imaging, the Z coordinate (tip height) is regulated by the feedback loop so as to maintain a given cantilever resonance frequency. In the idealized case of perfect tracking, the frequency will never deviate from this setpoint. Rather than adjusting the driving frequency to keep the cantilever in resonance at a particular probe height, the probe height is adjusted to restore the desired resonance frequency. As long as the overall regulation works, the end result will be the same, regardless of whether one uses frequency deviation or phase change[20,21] as the feedback signal.

What differs, depending on the choice of feedback signal, is the transient response of the topography regulation loop. Typically, topography tracking is achieved using a regulator with PI(D) (proportional, integral and occasionally differential) action, i.e. the response h to an input signal x is similar to

$$h(t) = -Px(t) - I\int_{-\infty}^{t} x(t')\mathrm{d}t' - D\frac{\mathrm{d}}{\mathrm{d}t}x(t). \tag{11}$$

When one is using a PLL as shown in Fig. 11, its internal frequency feedback loop also responds to the phase error signal with a PI(D) characteristic. In the frequency domain, each integral term corresponds to an additional 90° phase shift of the response with respect to a disturbance on the input. The integral terms of the cascaded PLL and the topography feedback loop thus combine to a total of 180°. This large cumulative phase shift results in a tendency to feedback oscillations. Even in the absence of self-oscillation, the feedback loop can raise noise levels around its natural frequency.

Phase modulation AFM (PM-AFM) does not use a frequency regulation loop, thereby avoiding one integration (and the corresponding 90° phase shift) of the error signal derived from the phase detector. This reduces the overall phase shift and makes topography feedback more stable. In the experiments shown later in this chapter, phase feedback has proven very valuable in obtaining high-resolution topographic images. On the other hand, phase detection is less suited to spectroscopy, where the cantilever resonance frequency is not kept constant. Since topography feedback is disabled in spectroscopy mode, phase and frequency detection are equally stable, but the frequency shift is a directly interpretable and thus generally more desirable spectroscopic quantity.

If the feedback characteristics of the PLL are adjustable over a sufficiently wide range, it is possible to switch seamlessly between phase- and frequency-modulated AFM. In the case where the PLL frequency tracking is much faster than the tip height correction by the topography feedback loop, the momentaneous resonance frequency excursions are tracked, and one obtains FM-AFM. When the frequency regulation is slow compared to the topography regulation, the excitation frequency does not track the cantilever resonance. Furthermore, if the PLL feedback is dominated by proportional action, the frequency shift output signal in Fig. 11 degenerates into a phase shift output signal. By varying the magnitude of integral and proportional action, the overall feedback loop can be continuously adjusted from phase-modulation-like to frequency-modulation-like. In this light, FM-AFM and PM-AFM are just border cases of the same technique, with real experiments being conducted somewhere in between.

In the application examples shown below, this flexibility was routinely used to optimize the PLL response on the fly to the task at hand (imaging or spectroscopy). Considering the stability advantages of PM-AFM and that FM-AFM and PM-AFM differ mainly in the choice of feedback parameters, it appears possible that some other "FM"-AFM imaging work may be — knowingly or unknowingly — closer to PM-AFM as well.

7.3. Application Examples

7.3.1. *Molecular resolution imaging of self-assembled monolayers*

Molecular and atomic resolution are the ultimate goal for high-resolution AFM techniques. Unfortunately, there are different interpretations of what the term "resolution" means in this context. Even with a relatively blunt

AFM tip, interacting with several atoms/molecules simultaneously, it is possible to see molecular or atomic structure by effectively averaging over an ensemble of molecules arranged in a lattice. A telltale characteristic of this "lattice resolution" is the inability to resolve molecular or atomic defects (Fig. 12). On the other hand, there is "true" molecular or atomic resolution, where *individual* molecules or atoms are sensed by the AFM probe. True molecular or atomic resolution is therefore usually demonstrated by imaging defect sites.

Figure 13 shows FM-AFM topography images of self-assembled alkanethiol monolayers on gold(111), immersed in an inert silicone oil (octamethylcyclotetrasiloxane). The lattice constant of the hexagonally packed organic molecules is on the order of 0.5 nm, which is not far from the

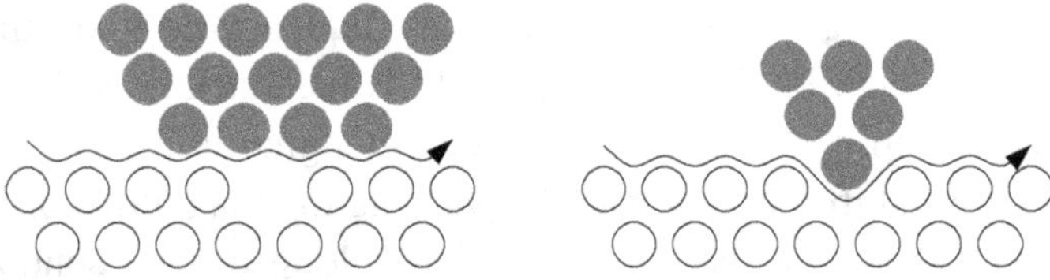

Fig. 12. Left: The scanning trajectory of a blunt tip can show periodic, averaged molecular detail, but is not capable of resolving individual molecules or defects. Right: Atomically sharp asperities are necessary for "true" atomic or molecular resolution.

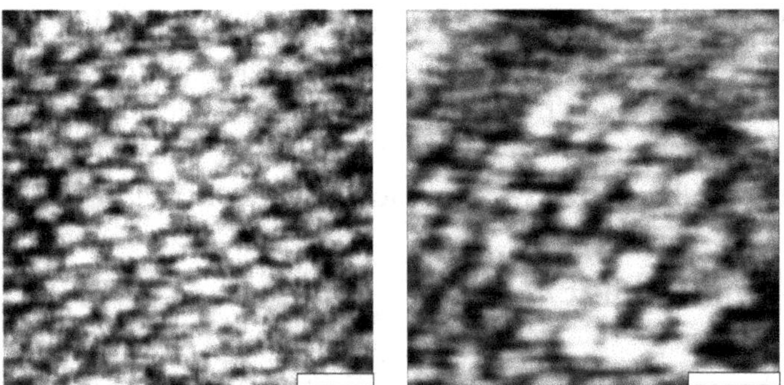

Fig. 13. Molecularly resolved self-assembled monolayers of 1-mercaptodecane/1-mercaptohexane codeposited on a gold(111) substrate immersed in OMCTS. The lateral resolution exceeds the lattice constant. Right: The resolved edge between a molecular island and the bare substrate indicates that true molecular resolution has been achieved. Taller (brighter) spots in the image are attributed to the longer-chain moiety embedded in a matrix of shorter-chain molecules. (Scale bar: 1 nm.)

atomic packing in some metals or semiconductors. From the lateral detail in these images, the resolution is estimated to be on the order of $\approx$0.2 nm, equivalent to atomic resolution. The peak–peak vibration amplitude was on a similar scale ($\approx$0.3 nm), at a positive frequency shift of about 5 kHz.

7.3.2. *Spectroscopy and structure of the liquid–solid interface*

The structure of liquids may change dramatically at the liquid–solid interface, be it by chemophysical interactions or simply as a result of the boundary conditions imposed by the solid surface. AFM techniques are uniquely suited for probing the structure of these interfacial regions, with the caveat that the structure of the liquid may not only be due to the interaction of the liquid with the solid substrate, but possibly also to the additional geometric confinement introduced by the AFM tip itself.

Oscillatory solvation forces are known to arise from the geometric confinement of liquid molecules between a solid substrate and the AFM tip (see also the chapter on conducting AFM in liquids). Contact AFM uses relatively soft cantilevers to achieve the sensitivity for measuring small solvation forces. However, while the solvation forces themselves may be weak, their small periodicity (given by the thickness of the molecular layers) results in large gradients. The deflection of soft cantilevers is unstable if the restoring cantilever spring constant k_{c} is outweighed by an attractive interaction force gradient k_{i}, i.e. $k_{\mathrm{c}} + k_{\mathrm{i}} < 0$. This "snap-in" instability can render attractive regions of the force spectrum experimentally inaccessible.

Modeling the cantilever with a lumped effective mass m, small-amplitude FM-AFM directly senses the force gradient via the frequency shift Δf,

$$(f_0 + \Delta f)^2 = \frac{k_{\mathrm{c}} + k_{\mathrm{i}}}{4\pi^2 m} = f_0 + \frac{k_{\mathrm{i}}}{4\pi^2 m}, \qquad (12)$$

which, in the limit of small Δf, reduces further to the linear relation

$$\Delta f \approx f_0 \frac{k_{\mathrm{i}}}{2k_{\mathrm{c}}}. \qquad (13)$$

Since the frequency shift scales with $k_{\mathrm{i}}/k_{\mathrm{c}}$, the cantilever spring constant k_{c} can be scaled so that $k_{\mathrm{c}} > |k_{\mathrm{i}}|$, avoiding the snap-in instability without unduly sacrificing sensitivity. This makes FM-AFM a particularly sensitive method for the study of such systems.

7.3.2.1. *Crystalline structure of n-dodecanol on graphite*

Some alkanes and alkane derivatives are known to layer on graphite substrates. However, the structure of these layers — beyond the first monolayer on the solid substrate — has been somewhat elusive. Scanning tunneling microscopy has been used to obtain high-resolution images of the first and second layers of the adsorbed liquid molecules.[22–24] Characterization of higher layers by STM is inhibited by the exponential decay of the tunneling current between tip and substrate as the tip height (gap size) is increased.

FM-AFM provides resolution that comes close to STM, yet does not rely on interacting with the substrate and lends itself to characterizing the structure of the liquid–solid interface further away from the substrate. Imaging of a second layer of an alkane using FM-AFM was recently reported.[25] In the following, a linear alcohol, n-dodecanol, was used as a layering liquid. The melting point of n-dodecanol is about 23°C, and experiments were performed at about 25°C. The substrate was freshly cleaved HOPG (highly ordered pyrolytic graphite). Stiff cantilevers ($k_c \approx 40\,\text{N/m}$) were used, with resonance frequencies on the order of 50 kHz in liquid. Vibration amplitudes were fixed at $A_{\text{pp}} \approx 0.25\,\text{nm}$ (peak-to-peak).

A high-resolution topographic image of a dodecanol layer is shown in Fig. 14. The linear molecules are seen to arrange side by side in parallel bands, and some submolecular features — most likely the polar OH groups — are evident.

Figure 15 shows a typical frequency shift spectrum as the tip height is varied. Note that the tip–sample separation scales shown are not absolute. Finding the origin (zero distance) would require one to apply a significant load to the probe, resulting in uncontrollable changes to the tip. For the same reason, it cannot be said with certainty whether Fig. 14 represents the first layer on the solid substrate or shows already a higher-order layer.

The frequency shift indicates pronounced oscillations that can be attributed to the layered structure of the molecules near the substrate. By

Fig. 14. $20 \times 20\,\text{nm}^2$ constant frequency topographic image of an n-dodecanol layer on graphite. ($\Delta f \approx +5\,\text{kHz}$).

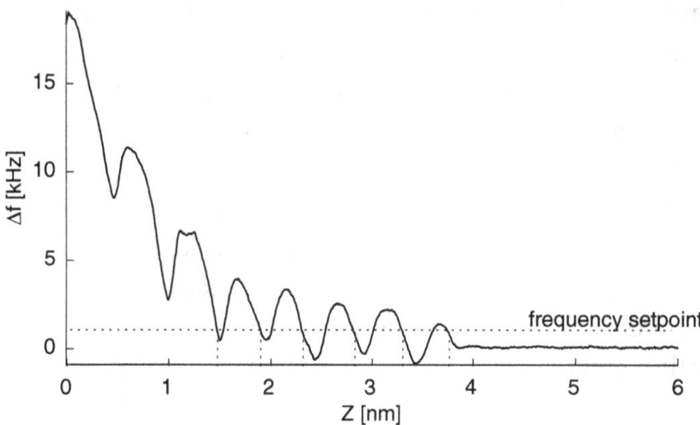

Fig. 15. Frequency shift (Δf) spectrum of dodecanol on HOPG. The modulation caused by solvation forces results in a nonmonotonic frequency vs. tip height relationship. For certain setpoints in frequency feedback imaging, the topography may be ambiguous among several layers (indicated), resulting in hysteresis.

choosing a suitable frequency setpoint, it should be possible to control the tip height to track the topography of different layers. Note that more than one layer may be able to sustain a given frequency shift. This results in hysteresis: the topography feedback may follow the current layer but, in the

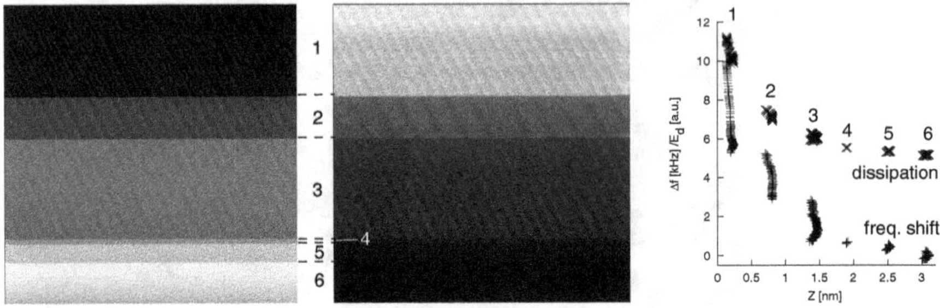

Fig. 16. Simultaneous topography (left) and dissipation (center) image ($100 \times 100\,\text{nm}^2$) of dodecanol on HOPG obtained while slowly varying the frequency shift setpoint Δf. The variable setpoint induces jumps between six consecutive layers, each of which can be stably imaged and reveals lateral molecular structure. Right: Correlations of tip height Z vs. frequency shift and per-cycle energy dissipation E_d.

presence of noise, it may "jump" to another layer capable of supporting the same frequency shift, making stable imaging difficult.

To demonstrate that imaging of several layers is possible, the frequency shift setpoint was slowly varied during the acquisition of a topographic image. This resulted in successive tracking of different layers (Fig. 16). The cross-section of the topographic image clearly shows discrete steps in the tip height of $\approx$0.5 nm, reflecting the thickness of the molecular layers.

An important feature of the topography image is that similar lateral structure persists in all the observed layers. The combination of lateral and vertical structures implies that the "liquid" solvation layers actually form a well-ordered, crystalline region up to several nanometers away from the supporting graphite substrate — at a temperature well above the bulk melting point.

Another implication is that true molecular resolution might not be achievable in such systems. From a practical point of view, crystal defects may not be stable enough to be imaged in this semifluid system, and thus true molecular resolution may be extremely difficult to demonstrate. More fundamentally, the general premise for achieving true molecular resolution is that a single atomic scale asperity provides the dominant interaction with the flat sample surface. However, when immersed in the layered interface region, the entire tip surface rather than a single asperity is in contact with successive layers, preventing the addressing of individual molecules.

7.3.2.2. *Dissipation*

Even without further instrumentation, one immediate by-product of the amplitude regulation loop in constant-amplitude FM-AFM is the amplitude of the driving force necessary for maintaining the oscillation. It is intuitively clear that this parameter is closely linked to the dissipation of the oscillating probe, as the mechanical energy transferred to the cantilever must exactly compensate for the energy losses in order to maintain the steady-state amplitude.

As the amplitude regulation signal is always present, it can also be recorded in parallel with the topography feedback as an alternative channel (Figs. 16 and 17 center), providing a second and independent contrast mechanism for free. It is also possible to use the drive amplitude (instead of the frequency shift) as a feedback signal. This would provide the equivalent of amplitude-detected AFM, with the important twist that the frequency regulation eliminates some effects of a nonlinear resonance curve, removing some of the instabilities of conventional, fixed-frequency AM-AFM.

However, it should be cautioned that even in a frequency feedback mode in a liquid environment, there is an inherent mechanism for crosstalk

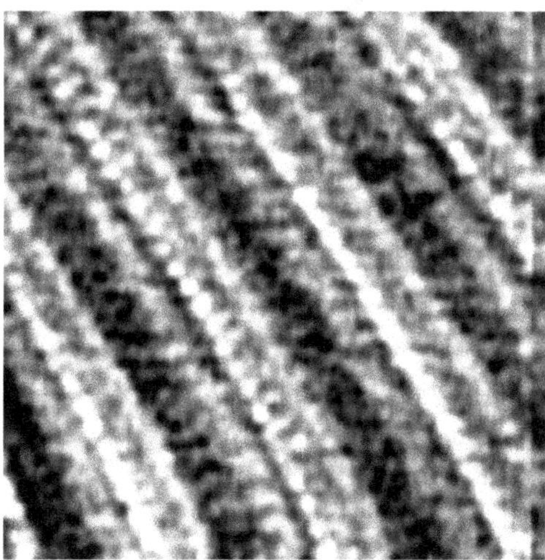

Fig. 17. Dissipation map acquired simultaneously with the topography data in Fig. 14. Brighter regions correspond to increased dissipation. Notice the partial contrast inversion, which is attributed to the crosstalk mechanism shown in Fig. 18.

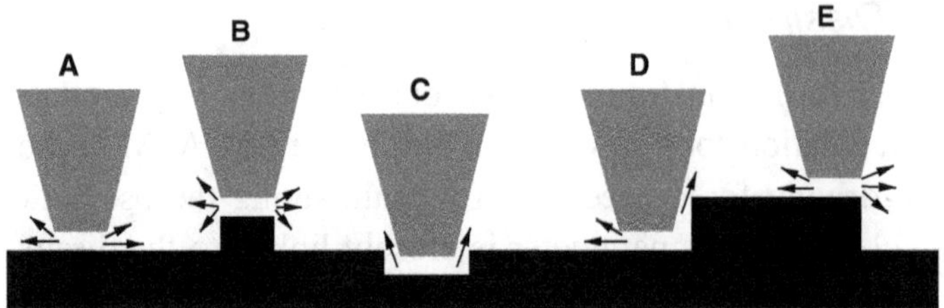

Fig. 18. Mechanism for crosstalk between topography and dissipation images. Topography features restrict or expand the space available for displacement of the liquid media by the oscillating tip, resulting in different amounts of viscous friction. A — normal case; B — reduced dissipation on an asperity; C — increased dissipation in a hole; D — increased dissipation at the bottom of a step edge; E — reduced dissipation at the top of a step edge.

between topography and dissipation. Dissipation arises not only from the tip–sample interaction but also from the flow of the liquid medium around the tip. In small "valleys," the liquid molecules are more strongly confined than on a flat area or on a small protrusion (Fig. 18). The varying degrees of confinement affect the magnitude of the hydrodynamic dissipation. Generally, it was observed that the dissipation will appear somewhat reduced on the higher side of a topography step, and increased on the lower side, in line with this simple model.

For a sinusoidal oscillation,[c] the mechanical work performed by an equally sinusoidal excitation force over one vibration cycle is

$$E_\mathrm{d} = \int_0^{2\pi} \frac{A_\mathrm{pp}}{2} \cos\psi \cdot \frac{F_\mathrm{pp}}{2} \sin(\psi + \phi)\mathrm{d}\psi = \frac{\pi A_\mathrm{pp} F_\mathrm{pp}}{4} \cdot \sin\phi, \qquad (14)$$

where A_pp and F_pp denote the peak-to-peak vibration and force amplitudes, and ψ is the phase shift between the cantilever deflection and the applied force. If the cantilever is kept at resonance (defined by a 90° phase shift), $\sin\phi = 1$, and the amplitude of the exciting force is directly proportional to the dissipation. In the case of magnetic excitation, the force in turn is directly proportional to the easily measured current through the driving solenoid, and it is therefore straightforward to quantify the dissipation.

[c]Distortions caused by nonlinear interactions are increasingly suppressed when reducing the vibration amplitude.

If the excitation force amplitude has not been directly calibrated (such as by measuring quasi-static cantilever deflections), the constant-excitation cantilever resonance curve,

$$A_{\mathrm{pp},0}(f) = \frac{F_{\mathrm{pp},0} f_0^2 k_{\mathrm{c}}}{4\pi^2 \sqrt{\left(f_0^2 - f^2\right)^2 + \left(\frac{f_0 f}{Q_0}\right)^2}}, \tag{15}$$

can be modeled to obtain the excitation force $F_{\mathrm{pp},0}$ and the quality factor Q_0 (the index 0 denotes that the parameters are obtained in the absence of tip–sample interaction). As the energy dissipation is directly described by the quality factor [Eq. (1)], one can directly calculate the energy loss per cycle in constant-amplitude mode:

$$E_{\mathrm{d}} = \frac{\pi k_{\mathrm{c}} A_{\mathrm{pp}}^2}{4 Q_0} \cdot \frac{F_{\mathrm{pp}}}{F_{\mathrm{pp},0}}. \tag{16}$$

Equation (16) depends only on the ratio of excitation forces (which can be substituted by the ratio of magnetic drive currents) and eliminates the need to calibrate the absolute drive force, as long as the quality factor Q_0 is known. This method is also applicable when the vibration mode of the cantilever differs from the quasi-static deflection, such as when using higher harmonics (see Fig. 5).

Figure 19 shows simultaneously acquired frequency shift and dissipation spectra of the dodecanol/graphite system, obtained with a tip that was somewhat blunted by extended use. The spectral features of both channels appear closely correlated (both reflecting the layered structure of the interface region). It is also apparent in both signals that the layered order does not decay gradually with increasing tip–sample distance, but that there is a sudden onset of the layering signature reflecting a sharp phase boundary between the crystalline region and the bulk liquid.

The dissipation can be decomposed into a baseline energy loss caused by the viscous damping of the cantilever immersed in the bulk liquid, and in an incremental dissipation attributed to the interaction of the tip with the layered interface region. The incremental dissipation associated with the outermost crystalline layer is not far from the thermal energy $k_{\mathrm{B}}T$. This could explain the nonexistence of further layers, which would be too weakly bound with respect to thermal activation. However, the crystallization from

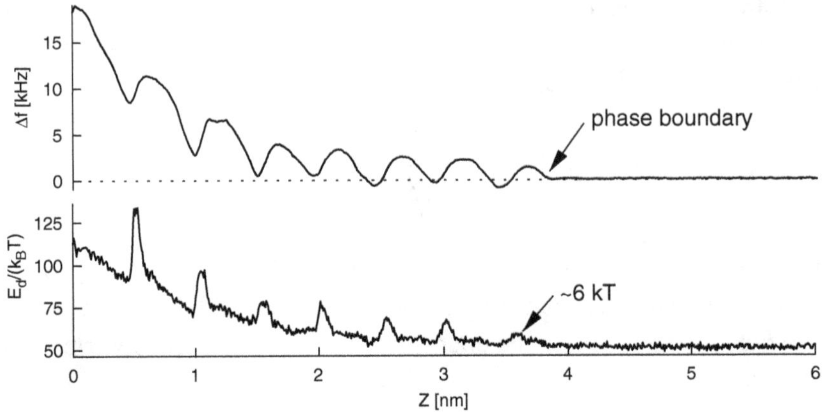

Fig. 19. Simultaneous frequency shift (Δf) and dissipation (E_d) spectra of dodecanol on HOPG. A sharp onset of layering suggests a defined phase boundary. The dissipation peak associated with squeezing out the top layer ($\Delta E_d \approx 6k_B T$) is marked.

the bulk liquid into the layers (or the expulsion by the AFM tip) is a collective process which cannot be explained in terms of a single molecular event, which makes more quantitative explanations difficult.

A second prominent feature is that the dissipation signal exhibits distinct narrow peaks. These peaks appear at tip–sample separations where the frequency shift decreases, reflecting diminished dynamic stiffness of the sample. This characteristic is intuitively understandable from the solid crystalline model of the dodecanol layers: a solid is characterized by an elastic response, i.e. increased sample stiffness and low dissipation during a compression–expansion cycle. When the pressure of the tip increases to the point of yield stress of the molecular layer, the crystalline structure is disrupted, and molecules are squeezed out from beneath the tip in a liquid-like fashion (Fig. 20). The layer recrystallizes when the tip is retreating again during the second half of the oscillation cycle. The repeated breakup of the crystalline order, as well as the viscous liquid-like flow of the squeezed-out molecules, dissipates energy.

The existence of nonmonotonic dissipation features has been discussed in the literature without an apparent consensus.[26–29] A recurrent criticism is that apparent dissipation features may be artefacts resulting from systematic errors in resonance frequency tracking. For example, an error in

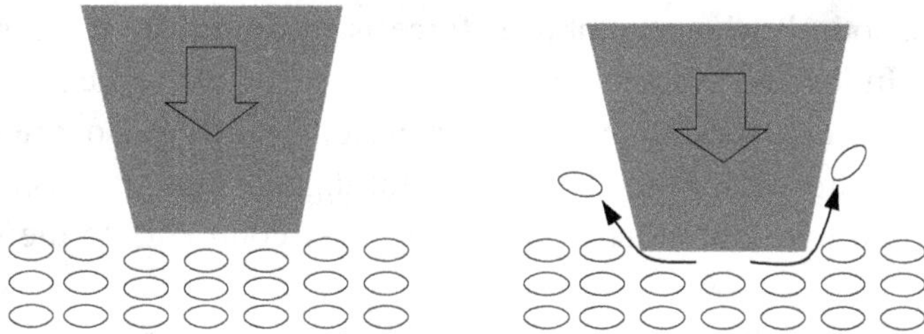

Fig. 20. Possible dissipation mechanism for small-amplitude vibrations in layered dodecanol. Left: Reversible, elastic compression of the solid layers results in high stiffness, but low dissipation. Right: During layer penetration, molecular layers liquefy to be repeatedly expelled/reassembled in a thermodynamically irreversible process, relaxing mechanical stress while simultaneously increasing dissipation.

the phase adjustment ($\phi \neq 90°$) would lead to intermixture of frequency shift and dissipation signals.[30,31] A phase error would also occur if parasitic resonances are present, and even if the resonance frequency was tracked accurately, the amplitude distortion imparted by the parasitics would induce apparent changes in dissipation whenever the frequency changes. All these effects can be ruled out here based on experimental evidence (the sharp dissipation features have no equivalent in the frequency shift spectrum), and the great care taken in implementing the magnetic cantilever actuation.

7.3.2.3. *Role of tip shape*

As noted above, the spectra in Fig. 19 were obtained with a blunted tip. For fresh/sharp tips, the dissipation oscillations were much less pronounced, or not noticeable at all, but could be induced by deliberate wear of the tip (such as by scanning in contact mode). This observation may help reconcile the conflicting reports in the literature. Possible explanations are that (i) sharp tips do not provide a single dominant gap size, thus blurring the dissipation spectra, and that (ii) blunt tips constrain the paths of expelled molecules much more strongly, enhancing viscous contributions to the observed dissipation.[32]

It also has been mentioned before that the nanoscale shape of the tip is crucial for obtaining true versus lattice resolution. The existence of multiple asperities with similar height at the end of the tip results in several "nanotips" whose combined force is measured by the cantilever dynamics. In layered systems, molecular order extends vertically. Therefore, in contrast to working on a two-dimensional surface, asperities sense ordered structures even if their heights are not aligned. This greatly enhances the likelihood that more than one asperity may contribute to the spectra (Fig. 21).

As there is a distinct phase boundary between the ordered (layered) and nonordered phases in the dodecanol–graphite system, the effect of vertically displaced asperities can be readily demonstrated in the frequency shift spectra. Figure 22 shows spectra obtained with a tip with multiple asperities. On approaching the surface, one asperity enters the layered region first

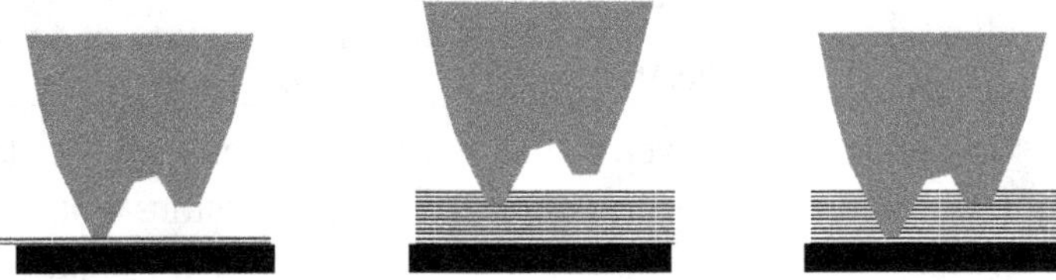

Fig. 21. Left: Recessed asperities do not interact with a two-dimensional, flat sample. Center and right: On tip approach, vertically staggered asperities are immersed successively in the three-dimensional solid–liquid interface region, resulting in the simultaneous sensing of different layers.

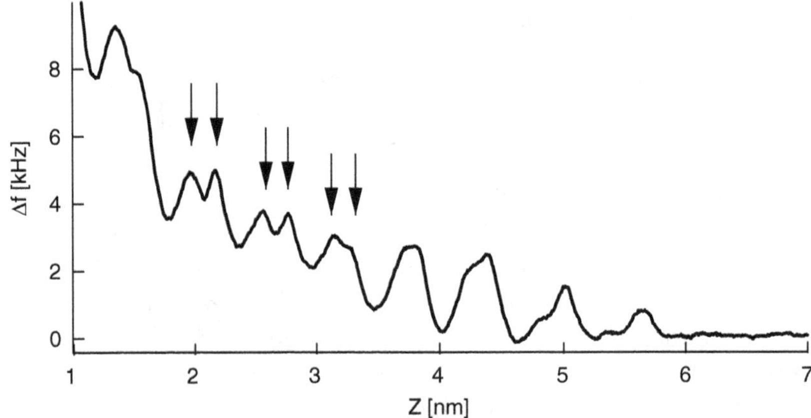

Fig. 22. Frequency shift spectrum of layered dodecanol, exhibiting split peaks attributed to vertically staggered asperities of the AFM tip.

(giving rise to periodic spectral features); only on further approach does a second asperity start interacting with the molecular layers and superimpose another set of spectral features. To fully resolve the ambiguities introduced by this type of artefact, the detailed shape of the AFM tip needs to be known. Conversely, a well-characterized and easily reproducible layered sample could be used for limited characterization of the 3D tip shape. However, full characterization of the atomic scale tip shape in a truly practical way remains one of the big unsolved problems for AFM.

7.4. Outlook: From Simple Organics to Biology

Scanning force microscopy in liquid media is a natural match for the study of biological samples due to their aqueous, electrically conductive native environment. AFM imaging of wet samples is typically applied to characterize relatively large biological structures, such as cell organelles, or large supramolecular assemblies. The field of view in such applications is often in the tens of micrometers. To a lesser extent AFM has been used to look at molecular-level biological samples, such as lipid membranes, proteins and DNA. At the mesoscopic level, AFM is a popular tool for example, observe phase separation in heterogeneous lipid membranes.

Spectroscopy-wise, AFM is common for mechanical characterization of cells (such as stiffness and cell adhesion). At the molecular level, AFM probes are used in semidestructive testing such as measuring antibody binding forces or unzipping DNA.

To realize the high-resolution potential of dynamic AFM, the systems under study have to be sufficiently solid-like and rigid to maintain a defined structure on the timescale of the AFM experiment. At the same time, one has to keep in mind that AFM is a technique for characterizing interfaces. The corresponding biological structures are lipid bilayer membranes, which are also where many fundamental biochemical processes happen.

The structure of the self-assembled organic monolayers shown in Fig. 13 is not far from that of a (supported) lipid membrane, showing the potential of FM-AFM for molecular biological research. Beyond the structure of the membranes themselves, biomembranes also serve as natural host matrices for "active" molecular assemblies such as proteins. If the membrane environment is sufficiently close to native conditions, high-resolution dynamic AFM may be able to observe individual biomolecular events, such as structural changes to proteins upon substrate binding. Observing individual specimens, rather than ensemble averages as in many other techniques, has the potential to provide a much more detailed and realistic picture of membrane-associated biological processes.

Biological structure and activity are closely linked to the aqueous, buffered environment. Membrane proteins fine-tune their conformation by hydrophilic–hydrophobic interactions within and outside of the membrane. As an example, liquid layering as observed in simple organics may also be relevant to biology. There is already some evidence that water undergoes layering next to a solid-supported lipid bilayer.[33–35] Such ordering would change the fluid properties of water, affecting the diffusion of molecules and ions to or from the membrane, and have a profound influence on membrane-associated biochemical equilibria and reaction rates. Furthermore, it may contribute to stabilizing the membranes themselves. This emphasizes the importance of providing a native or near-native environment.

The recent advances in dynamic force microscopy in a liquid environment, namely the vastly enhanced resolution, are therefore undoubtedly a great step toward bringing advanced scanning probe techniques from specialized physical laboratories to the interdisciplinary world of molecular-level biology.

References

1. F. J. Giessibl, *Rev. Mod. Phys.* **75**, 949 (2003).
2. F. J. Giessibl, *Mater. Today* **8**, 32 (2005).
3. T. Fukuma, K. Kobayashi, K. Matsushige and H. Yamada, *Appl. Phys. Lett.* **86**, 193108 (2005).
4. T. Fukuma, K. Kobayashi, K. Matsushige and H. Yamada, *Appl. Phys. Lett.* **87**, 034101 (2005).
5. M. Stark, R. W. Stark, W. M. Heckl and R. Guckenberger, *Proc. Nat. Acad. Sci.* **99**, 8473 (2002).
6. U. Dürig, *New J. Phys.* **2**, 5.1 (2000).

7. F. J. Giessibl, *Phys. Rev. B* **56**, 16010 (1997).
8. T. Fukuma, M. Kimura, K. Kobayashi, K. Matsushige and H. Yamada, *Rev. Sci. Instrum.* **76**, 053704 (2005).
9. T. Albrecht, P. Grütter, D. Horne and D. Rugar, *J. Appl. Phys.* **69**, 668 (1991).
10. T. Schäffer, J. Cleveland, F. Ohnesorge, D. Walters and P. Hansma, *J. Appl. Phys.* **80**, 3622 (1996).
11. U. Rabe, S. Hirsekorn, M. Reinstädtler, T. Sulzbach, C. Lehrer and W. Arnold, *Nanotechnology* **18**, 1 (2007).
12. B. Rogers, D. York, N. Whisman, M. Jones, K. Murray, J. Adams, T. Sulchek and S. Minne, *Rev. Sci. Instrum.* **73**, 3242 (2002).
13. S. Nishida, D. Kobayashi, T. Sakurada, T. Nakazawa, Y. Hoshi and H. Kawakatsu, *Rev. Sci. Instrum.* **79**, 123703 (2008).
14. S. Nishida, D. Kobayashi, H. Kawakatsu and Y. Nishimori, *J. Vac. Sci. Technol. B* **27**, 964 (2009).
15. H. Yamashita, N. Kodera, A. Miyagi, T. Uchihashi, D. Yamamoto and T. Ando, *Rev. Sci. Instrum.* **78**, 083702 (2007).
16. S. J. O'Shea, M. Welland and J. Pethica, *Chem. Phys. Lett.* **223**, 336 (1994).
17. W. Han, S. Lindsay and T. Jing, *Appl. Phys. Lett.* **69**, 4111 (1996).
18. S. de Beer, D. van den Ende and F. Mugele, *Appl. Phys. Lett.* **93**, 253106 (2008).
19. U. Dürig, O. Züger and A. Stalder, *J. Appl. Phys.* **72**, 1778 (1992).
20. T. Fukuma, J. I. Kilpatrick and S. P. Jarvis, *Rev. Sci. Instrum.* **77**, 123703 (2006).
21. H. Hölscher, *J. Appl. Phys.* **103**, 064317 (2008).
22. J. P. Rabe and S. Buchholz, *Science* **253**, 424 (1991).
23. L. C. Giancarlo and G. W. Flynn, *Annu. Rev. Phys. Chem.* **49**, 297 (1998).
24. G. Watel, F. Thiebaudau and J. Cousty, *Surf. Sci. Lett.* **281**, L297 (1993).
25. L. P. Van, V. Kyrylyuk, J. Polesel-Maris, F. Thoyer, C. Lubin and J. Cousty, *Langmuir* **25**, 639 (2009).
26. S. Patil, G. Matei, A. Oral and P. M. Hoffmann, *Langmuir* **22**, 6485 (2006).
27. S. J. O'Shea and M. Welland, *Langmuir* **14**, 4186 (1998).
28. T. Uchihashi, M. Higgins, Y. Nakayama, J. E. Sader and S. P. Jarvis, *Nanotechnology* **16**, S49 (2005).
29. G. B. Kaggwa, J. I. Kilpatrick, J. E. Sader and S. P. Jarvis, *Appl. Phys. Lett.* **93**, 011909 (2008).
30. J. E. Sader and S. P. Jarvis, *Phys. Rev. B* **74**, 195424 (2006).
31. S. J. O'Shea, *Phys. Rev. Lett.* **97**, 179601 (2006).
32. H.-W. Hu, G. A. Carson and S. Granick, *Phys. Rev. Lett.* **66**, 2758 (1991).
33. M. J. Higgins, M. Polcik, T. Fukuma, J. E. Sader, Y. Nakayama and S. P. Jarvis, *Biophys. J.* **91**, 2532 (2006).
34. T. Fukuma, M. Higgins and S. P. Jarvis, *Biophys. J.* **92**, 3603 (2007).
35. T. Fukuma, M. J. Higgins and S. P. Jarvis, *Phys. Rev. Lett.* **98**, 106101 (2007).

Wulf Hofbauer was awarded a doctorate of science (Dr. rer. nat.) in 2001 from the Technical University of Berlin for his work on experimental/methodological aspects of millimeter-wave electron magnetic resonance spectroscopy and its application for the study of membrane proteins. In his postdoctoral work at the National Biomedical Center for Advanced Electron Spin Resonance Technology (Cornell University), he developed a high-power, wide-band, quasi-optical magnetic resonance spectrometer for the investigation of protein dynamics. He joined IMRE in 2005 where he develops instrumentation and expertise for high-resolution dynamic force microscopy/spectroscopy in liquids.

Fabrication of Bio- and Nanopatterns by Dip Pen Nanolithography

Chapter 8

Qiyuan He*, Xiaozhu Zhou*, Freddy Y. C. Boey* and Hua Zhang*,†

Abstract

In this chapter, the recent research on fabrication of bio- and nanopatterns by dip pen nanolithography (DPN) is summarized. The DPN-generated bio- and nanopatterns include biomolecules (e.g. DNA, protein, enzyme, virus) and inorganic materials (e.g. nanoparticles, carbon nanotubes). DPN can also be used in electrochemistry and "click chemistry." In order to increase the throughput of DPN, 1D and 2D tip arrays and recent polymer pen lithography are developed.

8.1. Introduction

Nanoscience and nanotechnology has altered our understanding of the basic properties of materials with at least one spatial dimension in the size of 1–100 nm, resulting in different properties as compared to the bulk materials. Nanolithography is one of the core branches of nanoscience and nanotechnology, involving manipulation, etching, writing and printing at the nanometer scale.[1] Recently, considerable efforts have been made to

*School of Materials Science and Engineering, Nanyang Technological University, 50 Nanyang Avenue, Singapore 639798. Tel: (+65) 67905175; Fax: (+65) 67909081 http://www.ntu.edu.sg/home/hzhang

†Email: hzhang@ntu.edu.sg; hzhang166@yahoo.com

develop lithographic methods, owing to their scientific significance and technological applications in nanoelectronics, information storage, bioencoding and biodetection, etc.[2] Conventional lithography techniques have fulfilled many technological requirements. For example, electron beam lithography has been used to fabricate nanoelectrodes in nanoelectronics.[3] However, these conventional techniques are inherently limited with respect to either the resolution limit, or instrumental complexity or sophistication and inaccessibility, or incompatibility with biological building blocks.[4] In this regard, alternative approaches — such as scanning probe microscope (SPM)–based lithography, i.e. scanning probe lithography (SPL) — have been developed over the past years.[5] Of particular interest is SPL, since it offers both ultrahigh resolution/registration and *in situ* imaging capabilities. However, most of these methods suffer from a destructive nature and/or low throughput due to inherent limitations associated with the instruments.[6]

Dip pen nanolithography (DPN) is a departure from existing forms of SPL, allowing direct deposition of a wide variety of materials via a coated AFM tip onto surfaces of interest with nanometer resolution.[7–9] DPN has demonstrated its advantages over other SPL methods, as it is a direct-write and nondestructive lithographic technique, which allows routine patterning of nanostructures with high registration and resolution (e.g. sub-50 nm). Invented in 1999,[7] DPN has evolved into a full-fledged technique in various aspects. First, a myriad of "ink" molecules have been successfully patterned by DPN, including small organic molecules,[10,11] biological molecules,[12–14] polymers,[15,16] metal ions,[17,18] etc. Second, many kinds of substrates, including insulators, semiconductors and metals,[7–9,11,19–21] have been proven to be compatible with DPN. Most recently, the realization of parallel writing has rendered DPN capable of being a workhorse nanofabrication tool. One- and two-dimensional probe arrays with the number of probes up to 55,000 have been developed and have proven to be successful in the DPN process.[22] The parallelization of DPN definitely increases its throughput and will undoubtedly benefit many applications, such as large-area patterning of catalysts for growth of nanomaterials,[23] and high-throughput screening of cell–surface interactions.[24]

The previous DPN-related researches have been reviewed.[7,8] In this chapter, we will focus on the recent progress related to nano- and biopatterns generated on DPN-fabricated templates.

8.2. Biomolecules

Nanopatterns of biological molecules have great potential for addressing important fundamental questions in biological studies. DPN is extensively studied in patterning biological molecules with resolution limit orders of magnitude better than for conventional patterning techniques, such as photolithography. Many biomolecules, including proteins,[24–27] DNA,[12,28,29] peptides,[30–33] viruses[34–36] and bacteria,[37] are successfully patterned directly by DPN or indirect adsorption on DPN templates.

8.2.1. *DNA*

Patterning of DNA and oligonucleotides at the nanometer scale is important, due to their wide applications in fundamental biological studies. Demers *et al.* reported directly patterning modified oligonucleotides on metals and insulators by DPN.[12] The indirect patterning of DNA was also achieved by using patterns of gold nanostructure[28] and polymers[29] as templates. Despite the direct patterning of DNA molecules on target surfaces, several reports demonstrated facile assembly of other materials such as DNA-functionalized nanoparticles on DNA patterns.[38,39]

8.2.2. *Proteins*

Protein nanoarrays are extremely important in studying cell behaviors ranging from cell–cell adhesion to mechanisms of cell migration. The important progress was achieved by the Mirkin group in 2002.[24] Protein arrays were obtained by physical absorption of proteins on DPN-patterned 16-mercaptohexadecanioc acid (MHA) patterns. Lee *et al.* fabricated antibody nanoarrays by DPN to screen the human immunodeficiency antigen in serum samples. In this work, the antibodies were immobilized onto DPN-patterned MHA self-assembled monolayers (SAMs) to form antibody nanoarrays. More importantly, their result shows that the detection limit is more than 1000-fold higher than that of the conventional enzyme-linked immunosorbent assay (ELISA)–based immunoassys methods.[25] By using modified AFM tips,[13] they also reported direct writing of proteins on Au surface where cross-contamination can be avoided during the fabrication of complex multicomponent nanostructure assemblies.

A great number of works concerning protein nanoarrays are based on the covalent immobilization or even simple physisorption of proteins, which might suffer from several limitations, such as denaturation of proteins and/or unfavorable orientation. By combining DPN with molecular recognition-mediated self-assembly and molecular multivalency, Valiokas *et al.* achieved reversible assembly of histidine-tagged (His-tagged) proteins and performed a detailed study of protein–protein interactions.[40] Figure 1(a) shows a schematic representation of protein IFNRα2 and its interaction with quantum dots. Figures 1(b)–1(e) show fluorescent images of the reversible assembly of the target protein ifnar2-His_{10}.

Most recently, the Mirkin group introduced redox-activating dip pen nanolithography (RA-DPN)[27] to indirectly pattern oligonucleotides and

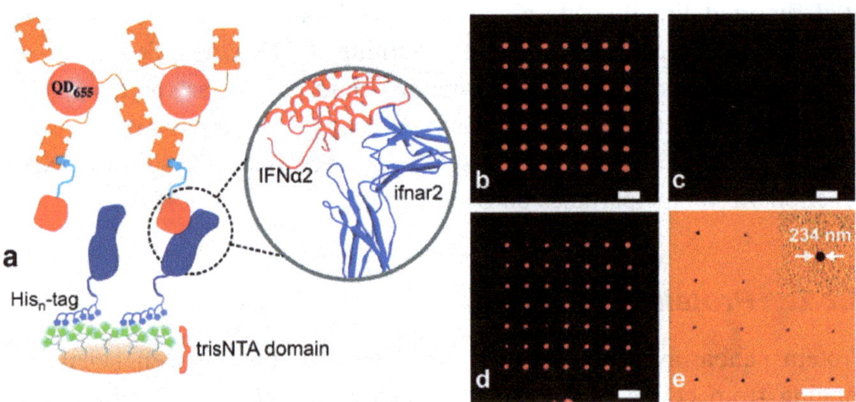

Fig. 1. (a) Schematic representation of the fluorescence assay for detection of the specific interaction between ifnar2-His_{10} and IFNRα2 in the trisNTA nanoarray (not to scale). In this assay, IFNRα2 is labeled with fluorescent quantum dots (QD_{655}). The close-up image is a fragment of a molecular model obtained by NMR analysis of the IFNRα2–ifnar2 complex. (b–d) Epifluorescence micrographs showing the functional activity of ifnar2-His_{10} receptors assembled on the trisNTA nanoarray using the assay described in panel a. (b) TrisNTA array after loading ifnar2-His_{10} and subsequently incubating it with IFNRα2-QD_{655} conjugate. (c) The same array after rinsing in 1 M imidazole. (d) The trisNTA array was reloaded with ifnar2-His_{10} and exposed to IFNRα2-QD_{655}. (e) Phase image obtained by tapping mode atomic force microscopy, showing a part of the same array after the removal of proteins by 100 mM HCl. The inset shows a close-up of one of the array spots. The scale bar in panels b–e is 3 μm. (*Reproduced with permission from Ref. 40.*)

proteins. The RA-DPN method relies on the delivery of the oxidant from an AFM tip to oxidize reduced hydroquinone (HQ) to the oxidized benzoquinone (BQ). The importance of this work is that it avoids the time-consuming optimization processes of patterning each additional species, since any material containing a strong nucleophile can be patterned through BQ templates.

8.2.3. *Enzymes*

Enzymes are attractive molecules because of their ability to efficiently catalyze specific reactions in mild conditions. Chilkoti and coworkers first reported the delivery of DNase I, a nonspecific endonuclease that digests double-stranded DNA into nucleotide fragments, onto the DNA-modified Au substrate by DPN.[41] The height measurement confirmed the digestion of the DNA monolayer, indicating the bioactivity of patterned enzymes. Besides direct deposition of enzymes, other applications have been demonstrated, i.e. use of an enzyme-modified AFM tip to catalyze surface reactions. Kaplan *et al.* reported peroxidase-catalyzed *in situ* polymerization of surface-oriented caffeic acid by DPN.[42] Willner *et al.* employed Au-nanoparticle-modified enzymes as biocatalyst to generate metallic nanowires.[43] In their work, glucose oxidase and alkaline phosphatase were used to selectively reduce Au and Ag salts to form Au and Ag nanowires on silica substrates.

8.2.4. *In situ growth of peptides*

Nanopatterning of peptides is of particular importance. It is easier to pattern peptides as compared to proteins. Meanwhile, peptides with a certain sequence have similar functions to proteins for cell–protein interactions. DPN has recently been extended to pattern peptides for various studies.[30–32] Many efforts have been made to directly pattern peptides with a specific combination of amino acids, but there are some limitations. On the one hand, new peptides will be required or synthesized for each new patterning. On the other, new experiment conditions need to be optimized for each new peptide patterning. Alternatively, our group recently developed a method for the controlled growth of peptides on SiO_x surface

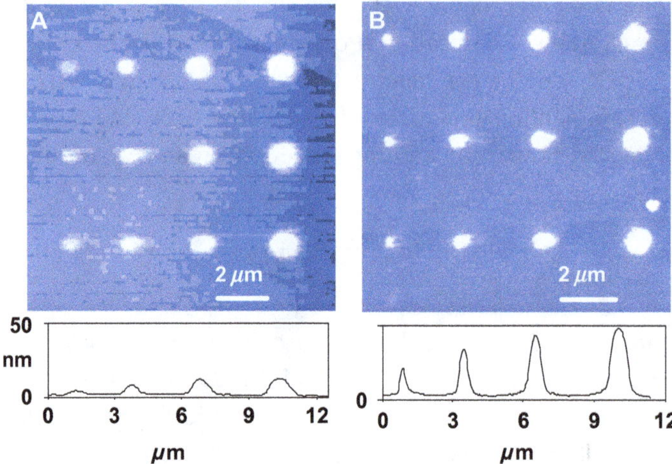

Fig. 2. AFM topography images and height profiles of (A) a DPN-generated PAMAM dot array on Si/SiO_x and (B) a peptide dot array subsequently grown on (A) after a 6 h ROP reaction. (*Copyright Wiley-VCH Verlag GmbH & Co. KGaA. Reproduced with permission from Ref. 33.*)

based on a combination of DPN and ring-opening polymerization (ROP) of tryptophan-N-carboxyanhydrides (Trp-NCAs).[33] The uniqueness of this method is that the DPN-generated amine-terminated polyamidoamine (PAMAM) dendrimer nanoarray serves as the anchoring scaffolds used for *in situ* growth of the peptide array. This is achieved by immersing the PAMAM patterned substrate in a Trp-NCA solution. Figure 2A shows the DPN-generated PAMAM dendrimer dot arrays on a Si/SiO_x substrate. After a 6 h ROP reaction, the height of the patterned dots was dramatically increased, indicating that the peptides successfully grew on the PAMAM dendrimers (Fig. 2B). Importantly, the height (i.e. chain length) of the synthesized peptides can be controlled by varying the ROP reaction time and concentration of the Trp-NCA in solution.

8.2.5. *Other biomolecules*

In addition, DPN appears to be a suitable technique for control of a single-virus particle, whose size is around 20–200 nm. Early work has shown that the DPN-generated chemical template can be used to immobilize a special modified cowpea mosaic virus.[34] Single-virus patterning was realized by Rafael *et al.* In their work, DPN-generated MHA patterns were first coordinated with Zn^{2+} ions. Then a single tobacco mosaic virus (TMS; tube shape — $300 \times 18\,nm^2$) was successfully immobilized on Au substrate with controlled orientation via metal binding to Zn^{2+}.[35] They also used an alternative approach to immobilization of TMS by antibody patterns. Since the feature size of this virus pattern is small enough, it can be applied to study single-cell infectivity.[36] Rozhok *et al.* reported patterning a single bacterium on *E. coli* antibodies/poly-L-lysine–modified MHA patterns.[37] They found that the biological activity of the bacterium remains for 4 h when attached to its antibodies.

8.3. Variant Possibility of DPN

A lot of DPN studies have been focused on patterning of small organic molecules, polymers and inorganic functional materials.[7–9] However, the applications of DPN are not limited to these materials. Here, we will present some recent developments of DPN used for patterning nanoparticles[41,42] and carbon nanotubes.[44,45]

8.3.1. *Nanoparticles*

Patterning of Au nanostructures at the submicrometer scale has attracted much attention recently, due to their unique electrical and optical properties. Our previous result shows that by using DPN-patterned MHA as etch resist, sub-50 nm Au dots can be generated with the wet chemical etching method.[46] Sukenik *et al.* employed microfluidic nanofountain probes to directly write Au nanowires consisting of Au nanocolloids.[47] Also, Au nanowires patterned by a biocatalyst-coated AFM tip were discussed above.[43] Bao and coworkers directly patterned Au NPs on silica substrate to form line and dot patterns by DPN.[47] In their work, aqueous solution of negatively charged citrate-capped Au nanoparticles (NPs) (5 nm in diameter), aqueous solution of positively charged amine-capped Au NPs (1.4 nm in diameter) and methanolic solution of positively charged 4-(*N*,*N*-dimethylamino) pyridine-capped Au NPs (average diameter of 5 nm) were used as inks to perform the direct writing of Au NPs. In order to achieve more homogeneous patterning, lubricating agents and several rounds of loading of Au NPs to the AFM tips were necessary to ensure the patterning quality, owing to the low-affinity nature of the inks.

Recently, the Mirkin group achieved direct assembly of Fe@C NPs with single-particle control using DPN-generated MHA templates.[48] In their work, a patterned MHA template with a feature size ranging from 50 to 120 nm was used to guide the assembly of Fe@C NPs via the hydrophilic interaction. Remarkably, a single NP array can be obtained with well-controlled placement and configuration.

8.3.2. *CNTs*

Creating arrays of carbon nanotubes (CNTs) with precise alignment and location at the nanometer scale has been essential to the intricate nano-electronics. However, creating such complex patterns is a challenge for researchers. Zou *et al.* aligned single-walled CNTs (SWCNTs) into ring structures through the van de Waals interaction between SWCNTs and COOH-terminated SAMs.[45] An alternative approach is the use of templates to control growth of nanomaterial patterns, such as CNTs. These templates contain a catalyst, which has been carefully patterned on substrates. Synthesis of the nanostructures is then performed on the patterned substrate. Recently, our group demonstrated the controlled growth of CNT patterns on DPN-generated catalyst templates.[23] We have developed a simple, efficient and uniform AFM tip-coating method, called the "scanning-coating" method, which is capable of patterning NP arrays over a large area without recoating the tip. The Co NP–coated tip can be used to generate Co NP dots with a feature size of less than 70 nm. Dots, lines, and even sophisticated patterns of Co NPs can be routinely generated and used as templates for controlled growth of CNTs. Figure 3A shows SWCNTs successfully grown on DPN-patterned Co catalyst dots on Si/SiO_x substrates. Furthermore, the direction of SWCNT arrays can be controlled when SWCNTs grow on the stable temperature-cut single-crystal quartz substrates, along the [100] crystallographic direction (Fig. 3B). In addition, DPN is capable of delivering Co NPs precisely to the desired location without contaminating other regions. This offers a convenient approach to observing the growth of SWCNTs, which has provided direct proof of the base growth mechanism for SWCNT formation in our experiment.

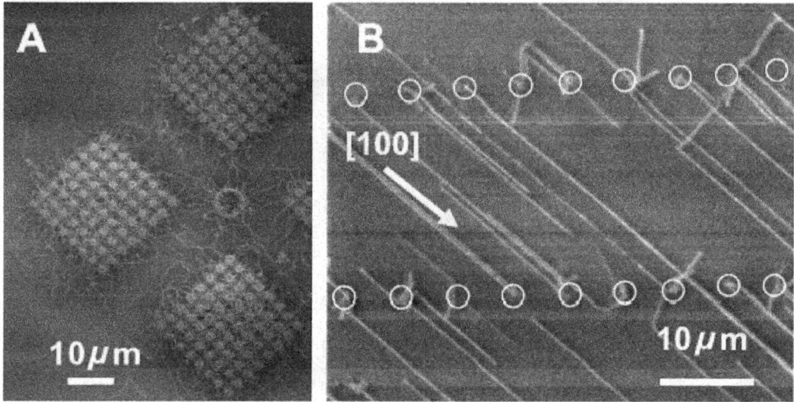

Fig. 3. SEM images of SWCNTs grown on the DPN-patterned Co NP catalyst dot array on Si/SiO_x (A) and quartz (B). (*Copyright Wiley-VCH Verlag GmbH & Co. KGaA. Reproduced with permission from Ref. 23.*)

8.4. Extension of DPN Capability

DPN has unlimited applications when combined with other techniques in various research fields. To demonstrate the power of DPN, we will present its several recent developments.

8.4.1. *Electrochemistry*

Li *et al.* demonstrated multiplex detection of two different target ssDNAs in solution using DPN-functionalized electrical gaps on the same chip.[49] The functionalization was achieved by direct deposition of ssDNA on electrical gaps to facilitate the capture of target ssDNAs in solution. With the help of DPN, a detection limit of 10 pM was reached.

8.4.2. *"Click" chemistry*

Frommer *et al.* applied DPN to "click" chemistry, a broad class of efficient coupling reactions, allowing a wide range of organic modifications to a surface at the submicrometer scale.[50] Specifically, a common Si_3N_4 AFM tip was inked with a catalyst and brought into contact with acetylide-modified Si substrate to perform the "click" reaction. Line and dot patterns were formed.

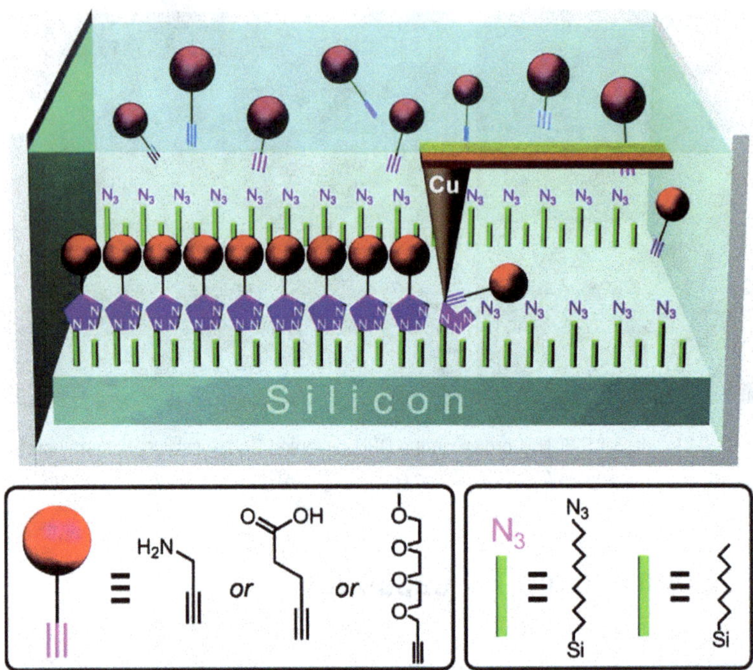

Fig. 4. Schematic representation of a scanning probe lithography approach to copper-catalyzed azide-alkyne cycloaddition (CuAAC) on surfaces, illustrating the three critical elements required for direct-write "click" chemistry: (1) a surface-anchored azide-terminated species that can be functionalized selectively with (2) a solvated terminal alkyne in the presence of (3) a copper catalyst. Terminal alkynes (including propargylamine, 4-pentynoic acid and 2,5,8,11-tetraoxatetradec-13-yne) dissolved in EtOH are "clicked" to a reactive surface, a mixed monolayer of 11-azido-undecyltrichlorosilane and octyltrichlorosilane (1:1) on a silicon wafer, as the copper-coated AFM tip is moved across the surface with the spatial resolution comparable to the tip dimensions. (*Reproduced with permission from Ref. 51.*)

Most recently, Walter *et al.* reported an alternative approach using Cu-coated tips to perform Cu-mediated azide-alkyne 1,3-dipolar cycloaddition (CuAAC) in solution, which resulted in a significantly higher degree of spatial control.[51] Figure 4 is a schematic representation of the surface "click" chemistry performed with Cu-coated AFM tips.

8.4.3. *Photomask*

Using DPN, the Mirkin group reported the fabrication of the positive-type photomask, which was used for interdigitated electrodes. The DPN patterning rate is 320 mm^2 min^{-1}, making it quite comparable to industrial standard methods based on lasers or electron beams. This application gives an example of exploration of DPN from research to practical industrial applications.[52]

8.4.4. *Modification of DPN probes*

The modification of AFM tips is an efficient way to explore the application of DPN or strengthen the experimental performance. Polydimethylsiloxane (PDMS) — coated AFM tips[53] have been proven to facilitate patterning by increased ink loading. Meanwhile, Espinosa *et al.* developed fountain pen nanolithography (FPN), a technique which employs a nanofountain probe tip with a microchannel structure to enable fluidic ink delivery.[54] FPN has found various applications in patterning silanes, proteins and other materials.

Most recently, a layer-by-layer (LbL) self-assembly technique was used to modify AFM tips. This porous polymeric structure provides a large ink reservoir for DPN experiments. Thus, a modified AFM tip gave an improved performance in patterning proteins by DPN.[55] Proteins were successfully patterned at the micrometer and submicrometer scales while retaining the biological activity for extended periods. Furthermore, the pore structures can provide a large amount of ink for longer-time and larger-area DPN experiments. The minimal feature size can be altered by changing either the composition of LbL SAMs or the environmental humidity during DPN. It is worth mentioning that in this DPN experiment, the feature size did not clearly follow the diffusion model, in which the tip–surface contact time is a critical factor. This result suggests a different mechanism for the ink transport, other than the classic diffusion model.[56]

8.5. Higher Throughput

8.5.1. *Parallel DPN*

The throughput limitation is a major problem that restricts the applications of DPN. Many studies have been carried out to improve the throughput of DPN for practical applications. For example, both linear[46,57] and 2D probe[58] arrays have been developed for massive production of homogeneous patterns. By employing the parallel DPN, Lenhert and coworkers succeeded in simultaneously delivering up to eight different functional lipids containing various amounts of biotin and/or nitrilotriacetic acid functional groups in a large area. Biomimetic membrane patterns were achieved by selective adsorption of functionalized or recombinant proteins based on the streptavidin or histidine-tag coupling, which is ideal for culturing cells.[59]

8.5.2. *Polymer pen lithography*

Most recently, the Mirkin group made a breakthrough in DPN with the introduction of polymer pen lithography (PPL),[60] a low-cost and cantilever-free lithographic approach to massive patterning. PPL employs elastomeric tips, made with a master prepared by conventional photolithography and subsequent wet chemical etching, as shown in Fig. 5. A maximum array of 11,000,000 pyramid-shaped tips were achieved in one 3×3 inch^2 wafer.

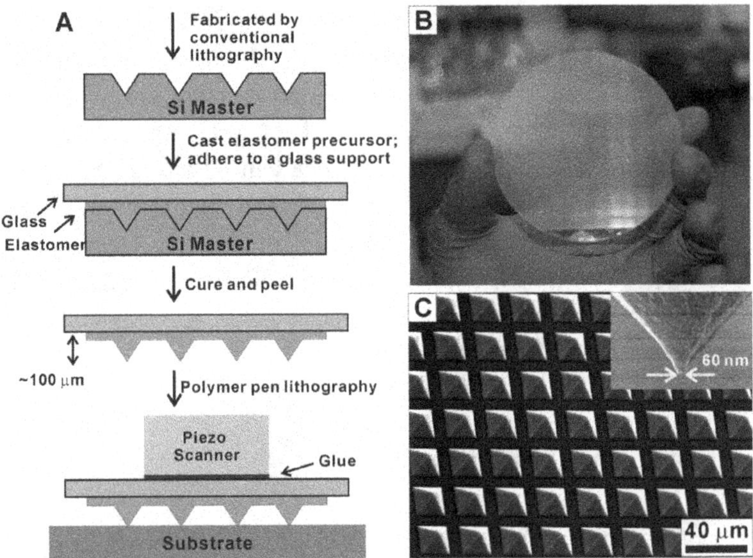

Fig. 5. (A) Schematic illustration of the polymer pen lithography setup. (B) Photograph of an 11-million-pen array. (C) Scanning electron microscope image of the polymer pen array. The average tip radius of curvature is 70 ± 10 nm (inset). (*Reproduced with permission from Ref. 60.*)

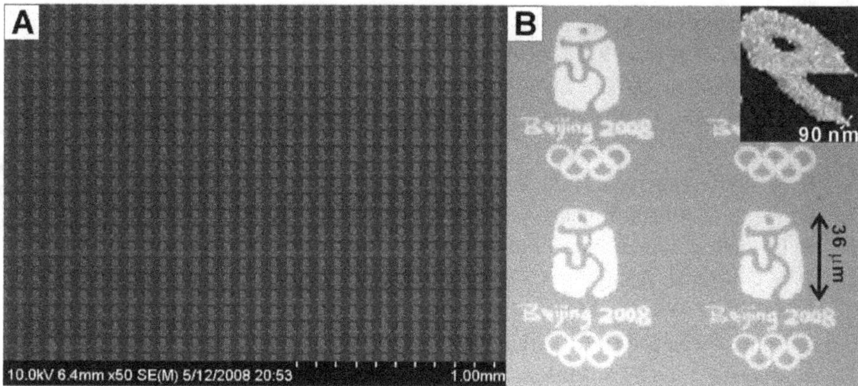

Fig. 6. SEM image of a representative region of ∼15,000 miniaturized duplicates of the 2008 Beijing Olympics logo. (B) Zoom-in optical image of a representative replica. The inset shows a magnified SEM image of the letter "e." (*Reproduced with permission from Ref. 60.*)

As proof of concept, the Mirkin group fabricated over 15,000 replicas of the 2008 Beijing Olympics logo (each contains 24,000 dots) in 40 min, with a yield of more than 99% in 1 cm^2 substrate, as shown in Fig. 6.

8.6. Conclusion

We have presented a brief overview of the recent progress of dip pen nanolithography, with a particular emphasis on the patterning of bio- and nanomolecules directly or indirectly. The development of DPN makes it a promising tool in fundamental research areas, such as biomolecule interactions, cell screening and infectivity. Of particular importance is the introduction of the parallel linear and 2D pens and polymer pen lithography, which resolve the inherent throughput limitations and thus render DPN possible for fast and large-area fabrications.

References

1. D. M. Eigler and E. K. Schweizer, *Nature* **344**, 524 (1990).
2. Y. Xia, J. A. Rogers, K. E. Paul and G. M. Whitesides, *Chem. Rev.* **99**, 1823 (1999).

3. C. Schönenberger, A. Bachtold, C. Strunk, J. P. Salvetat and L. Forró, *Appl. Phys. A: Mater. Sci. Process.* **69**, 283 (1999).
4. S. Zauscher, in *Dekker Encyclopedia of Nanoscience and Nanotechnology*, eds. J. A. Schwars, C. I. Contescu and K. Putyera (Taylor & Francis, LLC. 2004), pp. 2973–2983.
5. R. M. Nyffenegger and R. M. Penner, *Chem. Rev.* **97**, 1195 (1997).
6. G. Y. Liu, S. Xu and Y. Qian, *Acc. Chem. Res.* **33**, 457 (2000).
7. R. D. Piner, J. Zhu, F. Xu, S. Hong and C. A. Mirkin, *Science* **283**, 661–663 (1999).
8. D. S. Ginger, H. Zhang and C. A. Mirkin, *Angew. Chem. Int. Ed.* **43**, 30 (2004).
9. K. Salaita, Y. Wang and C. A. Mirkin, *Nat. Nanotechnol.* **2**, 145 (2007).
10. S. Hong, J. Zhu and C. A. Mirkin, *Science* **286**, 523 (1999).
11. A. Ivanisevic and C. A. Mirkin, *J. Am. Chem. Soc.* **123**, 7887 (2001).
12. L. M. Demers, D. S. Ginger, S. J. Park, Z. Li, S. W. Chung and C. A. Mirkin, *Science* **296**, 1836 (2002).
13. K. B. Lee, J. H. Lim and C. A. Mirkin, *J. Am. Chem. Soc.* **125**, 5588 (2003).
14. J.-H. Lim, D. S. Ginger, K.-B. Lee, J. Heo, J. M. Nam and C. A. Mirkin, *Angew. Chem. Int. Ed.* **42**, 2309 (2003).
15. J.-H. Lim and C. A. Mirkin, *Adv. Mater.* **14**, 1474 (2002).
16. A. Noy, A. E. Miller, J. E. Klare, B. L. Weeks, B. W. Woods and J. J. DeYoreo, *Nano Lett.* **2**, 109 (2002).
17. L. Ding, Y. Li, H. Chu, X. Li and J. Liu, *J. Phys. Chem. B* **109**, 22337 (2005).
18. J. Li, C. Lu, B. Maynor, S. Huang and J. Liu, *Chem. Mater.* **16**, 1633 (2004).
19. C. M. Bruinink, C. A. Nijhuis, M. Peter, B. Dordi, O. Crespo-Biel, T. Auletta, A. Mulder, H. Schonherr, G. J. Vancso, J. Huskens and D. N. Reinhoudt, *Chem. Eur. J.* **11**, 3988 (2005).
20. S. E. Kooi, L. A. Baker, P. E. Sheehan and L. J. Whitman, *Adv. Mater.* **16**, 1013 (2004).
21. G. Agarwal, R. R. Naik and M. O. Stone, *J. Am. Chem. Soc.* **125**, 7408 (2003).
22. K. Salaita, Y. Wang, J. Fragala, R. A. Vega, C. Liu and C. A. Mirkin, *Angew. Chem. Int. Ed.* **45**, 7220 (2006).
23. B. Li, C. F. Goh, X. Zhou, G. Lu, H. Tantang, Y. Chen, C. Xue, F. Y. C. Boey and H. Zhang, *Adv. Mater.* **20**, 4873 (2008).
24. K.-B. Lee, S.-J. Park, C. A. Mirkin, J. C. Smith and M. Mrksich, *Science* **295**, 1702 (2002).
25. K. B. Lee, E. Y. Kim, C. A. Mirkin and S. M. Wolinsky, *Nano Lett.* **4**, 1869 (2004).
26. H. Jung, C. K. Dalal, S. Kuntz, R. Shah and C. P. Collier, *Nano Lett.* **4**, 2171 (2004).
27. A. B. Braunschweig, A. J. Senesi and C. A. Mirkin, *J. Am. Chem. Soc.* **131**, 922 (2009).
28. H. Zhang, Z. Li and C. A. Mirkin, *Adv. Mater.* **14**, 1472 (2002).
29. D. Nyamjav and A. Ivanisevic, *Adv. Mater.* **15**, 1805 (2003).
30. Y. Cho and A. Ivanisevic, *Langmuir* **22**, 8670 (2006).
31. J. Auernheimer, C. Dahmen, U. Hersel, A. Bausch and H. Kessler, *J. Am. Chem. Soc.* **127**, 16107 (2005).
32. Y. Cho and A. Ivanisevic, *J. Phys. Chem. B* **108**, 15223 (2004).

33. X. Zhou, Y. Chen, B. Li, G. Lu, F. Y. C. Boey, J. Ma and H. Zhang, *Small* **4**, 1324 (2008).
34. C. L. Cheung, J. A. Camarero, B. W. Woods, T. Lin, J. E. Johnson and J. J. De Yoreo, *J. Am. Chem. Soc.* **125**, 6848 (2003).
35. R. A. Vega, D. Maspoch, K. Salaita and C. A. Mirkin, *Angew. Chem. Int. Ed.* **44**, 6013 (2005).
36. R. A. Vega, C. K.-F. Shen, D. Maspoch, J. G. Robach, R. A. Lamb and C. A. Mirkin, *Small* **3**, 1482 (2007).
37. S. Rozhok, C. K.-F. Shen, P.-L. H. Littler, Z. Fan, C. Liu, C. A. Mirkin and R. C. Holz, *Small* **1**, 445 (2005).
38. D. Nyamjav and A. Ivanisevic, *Biomaterials* **26**, 2749 (2005).
39. S.-W. Chung, D. S. Ginger, M. W. Morales, Z. Zhang, V. Chandrasekhar, M. A. Ratner and C. A. Mirkin, *Small* **1**, 64 (2005).
40. T. Rakickas, M. Gavutis, A. Reichel, J. Piehler, B. Liedberg and R. Valiokas, *Nano Lett.* **8**, 3369 (2008).
41. J. Hyun, J. Kim, S. L. Craig and A. Chilkoti, *J. Am. Chem. Soc.* **126**, 4770 (2004).
42. P. Xu, H. Uyama, J. E. Whitten, S. Kobayashi and D. L. Kaplan, *J. Am. Chem. Soc.* **127**, 11745 (2005).
43. B. Basnar, Y. Weizmann, Z. Cheglakov and I. Willner, *Adv. Mater.* **18**, 713 (2006).
44. Y. Wang, D. Maspoch, S. Zou, G. C. Schatz, R. E. Smalley and C. A. Mirkin, *Proc. Natl. Acad. Sci. U.S.A.* **103**, 2026 (2006).
45. S. Zou, D. Maspoch, Y. H. Wang, C. A. Mirkin and G. C. Schatz, *Nano Lett.* **7**, 276 (2007).
46. H. Zhang, N. A. Amro, S. Disawal, R. Elghanian, R. Shile and J. Fragala, *Small* **3**, 81 (2007).
47. H. Taha, A. Lewis and C. Sukenik, *Nano Lett.* **7**, 1883 (2007).
48. W. M. Wang, R. M. Stoltenberg, S. Liu and Z. Bao, *ACS Nano* **2**, 2135 (2008).
49. S. Li, S. Szegedi, E. Goluch and C. Liu, *Anal. Chem.* **80**, 5899 (2008).
50. D. A. Long, K. Unal, R. C. Pratt, M. Malkoch and J. Frommer, *Adv. Mater.* **19**, 4471 (2007).
51. W. F. Paxton, J. M. Spruell and J. F. Stoddart, *J. Am. Chem. Soc.* **131**, 6692 (2009).
52. J.-W. Jang, R. G. Sanedrin, A. J. Senesi, Z. Zheng, X. Chen, S. Hwang, L. Huang and C. A. Mirkin, *Small* **5**, 1850 (2009).
53. H. Zhang, R. Elghanian, N. A. Amro, S. Disawal and R. Eby, *Nano Lett.* **4**, 1649 (2004).
54. K.-H. Kim, N. Moldovan and H. D. Espinosa, *Small* **1**, 632 (2005).
55. C.-C. Wu, H. Xu, C. Otto, D. N. Reinhoudt, R. G. H. Lammertink, J. Huskens, V. Subramaniam and A. H. Velders, *J. Am. Chem. Soc.* **131**, 7526 (2009).
56. J. Jang, G. C. Schatz and M. A. Ratner, *J. Chem. Phys.* **116**, 3875 (2002).
57. K. Salaita, S. W. Lee, X. Wang, L. Huang, T. M. Dellinger, C. Liu and C. A. Mirkin, *Small* **1**, 940 (2005).

58. C. A. Mirkin, *ACS Nano* **1**, 79 (2007).
59. S. Sekula, J. Fuchs, S. Weg-Remers, P. Nagel, S. Schuppler, J. Fragala, N. Theilacker, M. Franzreb, C. Wingren, P. Ellmark, C. A. K. Borrebaeck, C. A. Mirkin, H. Fuchs and S. Lenhert, *Small* **4**, 1785 (2008).
60. F. Huo, Z. Zheng, G. Zheng, L. R. Giam, H. Zhang and C. A. Mirkin, *Science* **321**, 1658 (2008).

Qiyuan He obtained a BS in Applied Chemistry from Wuhan University in 2005, and an MS from Fudan University in 2008. He is currently pursuing his PhD under Prof. Hua Zhang at Nanyang Technological University.

Xiaozhu Zhou obtained a BS in Materials Science from Zhejiang University in 2006, and obtained a PhD under Profs. Freddy Boey and Hua Zhang at Nanyang Technological University in 2010. He is now working as a research fellow under Profs. Hua Zhang and Venkatraman Subbu at Nanyang Technological University.

Freddy Boey is presently Professor and Chair in the School of Materials Science and Engineering, Nanyang Technological University. He is Co-director of the new Nanyang Institute of Engineering in Medicine. He is also Director of Nanofrontier, an NTU incubator. He has founded or cofounded five startup companies, which have received investment funding. Under Prof. Boey 22 PhDs and 12 postdocs. He presently has a team of 3 Senior RFs, 10 postdocs and 16 PhDs. His main research focuses on biomedical cardiovascular/ocular drug-releasing implants, functional biomaterials, nanomaterials and nanostructures for cell regeneration, sensing and energy storage.

Hua Zhang obtained his BS and MS from Nanjing University in China in 1992 and 1995, respectively, and completed his PhD under Prof. Zhongfan Liu at Peking University in China in 1998. He joined Prof. Frans C. De Schryver's group at Katholieke Universiteit Leuven (KULeuven) in Belgium as a Research Associate in 1999. In 2001, he moved to Prof. Chad A. Mirkin's group at the Northwestern University in the USA as a Postdoctoral Fellow. He started to work at NanoInk Inc., USA, as a Research Scientist/Chemist in 2003. After that, he worked as a Senior Research Scientist at the Institute of Bioengineering and Nanotechnology in Singapore, from November 2005 to July 2006. Dr Zhang is now an Assistant Professor at Nanyang Technological University. He has published more than 20 patent applications and more than 90 papers, with an H-index of 24. He has been an Associate Editor of *International Journal of Nanoscience* since 2007, has been on the Editorial Board of *NANO* since the same year, and has been a member of the Advisory Committee of IOP Asia-Pacific since 2010. He has been invited to give keynote/invited talks at many international conferences, serve as session chairs, and has also organized several international conferences.

Atomic Force Microscopy-Based Nano-Oxidation

Chapter 9

Xian Ning Xie,*,§ Hong Jing Chung,† and Andrew T. S. Wee*,‡,¶

Abstract

This chapter reviews the nano-oxidation technique based on atomic force microscopy (AFM) for nanoscale patterning and fabrication. The mechanism, technical development and applications of AFM nano-oxidation are summarized. The materials for nano-oxidation, the aspect ratio of nano-oxide, and the chemical and dielectric properties of nano-oxide are discussed.

9.1. Introduction

Scanning probe microscopy (SPM) nanolithography is a novel method for nanoscale patterning and fabrication. It is based on two probe-related techniques: atomic force microscopy (AFM)[1] and scanning tunneling microscopy (STM).[2] There are numerous review articles[3–11] describing the various aspects of SPM nanolithography, and a comparison of SPM nanolithography with conventional methods is provided in Table 1 of Ref. 12. The similarities and differences between AFM and STM nanolithography

*NUS Nanoscience and Nanotechnology Initiative, National University of Singapore.
†University of Twente, The Netherlands.
‡Department of Physics, National University of Singapore, Singapore, 117542.
§Email: nnixxn@nus.edu.sg
¶Email: phyweets@nus.edu.sg

are presented in Table 2 of Ref. 12. In general, STM nanolithography has a higher resolution, but is limited to conducting and semiconducting samples, while AFM nanolithography is capable of patterning insulating materials. The two techniques complement each other, and their combination makes SPM nanolithography a versatile approach to nanopatterning a wide range of materials. This review emphasizes the nano-oxidation technique[3] in AFM nanolithography, and is structured as follows. We first briefly introduce the various AFM-based nanofabrication methods developed in the last two decades. We then focus on AFM nano-oxidation, which is one of the earliest and most-widely-studied techniques in AFM nanolithography. The oxide formation mechanism, technical development and applications of AFM nano-oxidation will also be discussed.

AFM nanolithography can be classified into two general groups in terms of their operational principles: (i) force-assisted nanolithography and (ii) bias-assisted nanolithography.[3] In force-assisted AFM nanolithography, a large force is loaded on the probe for pattern fabrication, and the tip–surface interaction is mainly mechanical. In bias-assisted AFM nanolithography, an electrical bias is applied to the probe to establish a strong field in the tip–surface nanojunction. The field strength is usually in the regime of 10^8 V/m to 10^{10} V/m, such that various physical and chemical processes can be initiated to facilitate pattern formation on the substrate surface. The technical specifications and typical examples of the two groups of techniques are summarized in Table 3 of Ref. 12, and more detailed descriptions of each technique are available in Ref. 3. Among the many nanolithographic methods, we highlight AFM nano-oxidation in this review. This is because nano-oxidation is the most representative method in bias-assisted AFM nanolithography, and it has wide applications in nanopatterning and nanodevice fabrication. In contrast to the well-understood thermal oxidation, AFM nano-oxidation is very complicated, and involves both chemical and physical processes, such as the ionic dissociation of water, the diffusion of space charges and the anodic reaction. In addition, the formation of a nanoscopic water meniscus bridging the tip–surface gap and the dependence of the meniscus on the tip–surface spacing, humidity condition and field strength render the control of nano-oxidation rather difficult. Moreover, static and transient charge effects influence the aspect ratio of nano-oxides. Due to this complexity, the results reported in the literature are rather diverse, and there is still much debate on the limiting factors in the formation of nano-oxides. Therefore, it is the aim of this review to provide a succinct summary of the key issues in AFM nano-oxidation.

9.2. Mechanism of Nano-oxidation

Figure 1 shows a schematic of AFM nano-oxidation. The AFM probe is placed above the semiconductor Si substrate with a tip–substrate spacing of <1 nm. A water meniscus is usually formed at the tip–surface gap due to capillary condensation at the sharp tip apex, and such condensation is greatly enhanced by the application of an electric field. When the tip is negatively biased, water molecules dissociate to O^- and OH^- ions, which are driven by the field into the Si substrate. The oxidative ions react with Si to form Si oxide, as described by the chemical reaction formula: $Si + 4OH^- + 4h \rightarrow SiO_2 + 2H_2O$. Due to the incorporation of H atoms and O vacancy defect formation, the chemical stoichiometry of nano-oxide deviates from SiO_2, and the use of SiO_x ($x \leq 2$) in the representation of AFM oxide is more relevant. The molecular volume of the oxide is usually larger than that of the substrate; for example, the volume of oxide is approximately two times that of substrate in the case of Si. Therefore, protruded oxide patterns such as dots and lines are generated on the substrate as a result of nano-oxidation, as shown in Fig. 1.

There are a number of models proposed for the mechanistic description of AFM nano-oxidation, and they are summarized in Table 1. In general,

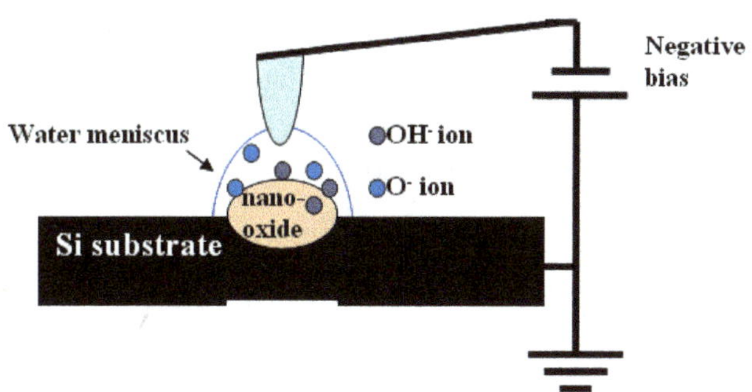

Fig. 1. Schematic of AFM nano-oxidation.

Table 1. Summary of models proposed for AFM nano-oxidation.

Nano-oxidation model	Description
Cabrera–Mott model	• Thickness of oxide is governed by diffusion-limited electric field.
Power law model	• AFM oxidation is observed only for voltages exceeding a doping-dependent threshold above which oxidation kinetics follows a power law.
Log-kinetic model	• Height (h) proportional to log $(1/V)$, $V^{-1/2}$ and $V^2 - 1/4$ and linear behavior between $1/h$ and log V.
Space charge model	• Varied space charge dependence of oxidation process as function of substrate doping type/level. • Space charge effects are consistent with rapid decline of high initial growth rates. • Alberty–Miller scheme to describe direct pathway for reaction of oxyanions with silicon at Si/SiO_2 interface.

the nano-oxidation mechanism and kinetics are closely related to the electrical field, surface stress, water meniscus formation and OH^- diffusion. The specific models include the Cabrera–Mott model,[13] power law model,[14] log-kinetic model[15–18] and space charge model,[19–22] each of which emphasizes a different aspect of nano-oxidation. These models are valid under specific experimental conditions, such as contact mode or tapping mode oxidation. Among them, the space charge model[19–22] is useful for explaining the formation of high-aspect-ratio oxide. This model involves space charge buildup at the oxide/substrate interface which generates an energy barrier and hinders the further diffusion of OH^- oxidants and the oxidation of the substrate. As will be discussed later, charge modulation using opposite tip bias polarity could minimize or remove the space charge for the fabrication of high-aspect-ratio oxide.

9.3. Materials Used in Nano-oxidation

Semiconductors and metals are the two major families of materials suitable for AFM nano-oxidation. These substrates are conductive, and thus allow bias application and the ionic dissociation of the water meniscus which is essential for the generation of OH^- oxidants in the nano-oxidation reaction. More importantly, the nano-oxidation of these materials is of great significance in nanotechnology, as the nano-oxide can serve as gate oxide on semiconductors and metals for nanodevice fabrication. The semiconductors subjected to nano-oxidation are Si, Si_3N_4, SiC, Ga[Al]As, etc.[23–36] Extensive investigations have also been carried out on the nano-oxidation of metals such as Ti, Al, Cr, Nb, Ni, Ta, Mo, Zr and Co.[37–50] In addition, there is interest in the nano-oxidation of molecularly functionalized/passivated surfaces.[51–58] The aim of such work is to explore the role of surface passivation in oxide formation, and to transfer patterns through chemical etching and molecular assembly. Readers are referred to Table 5 of Ref. 12 for a summary of the different materials used in AFM nano-oxidation, and the corresponding nanostructures fabricated. Recently Garcia and coworkers extended AFM nano-oxidation to produce nanostructures made of materials other than oxides.[59,60] Their strategy is based on the formation of organic menisci to tune the chemical composition of structures formed. The water meniscus is replaced by an organic meniscus, and the oxidation reaction is eliminated or significantly suppressed. Nanometer-sized menisci of organic liquids such as octane and 1-octene were used to confine chemical reactions.[59] Their results indicated that the composition of the fabricated structures is organic-solvent-dependent. The growth rate can be significantly modified by the composition of the organic solvent. They confirmed that AFM nano-oxidation is a general process that is applicable to vapor and liquid phases, as well as polar and nonpolar solvents.

9.4. Spreading Modes of OH^- Oxidants

In nano-oxidation, one critical factor is the lateral diffusion of OH^- oxidants, as it determines the resolution of the oxide structures formed. It is commonly accepted that the lateral dimension of oxides is related to the diffusion of OH^- oxidants. Some authors observed core/base oxide patterns where the base is ring-shaped, and the core is dome-shaped, formed directly under the tip apex.[61–63] The core oxide is attributed to the oxidation product at the early stage of bias application at which the oxidation is

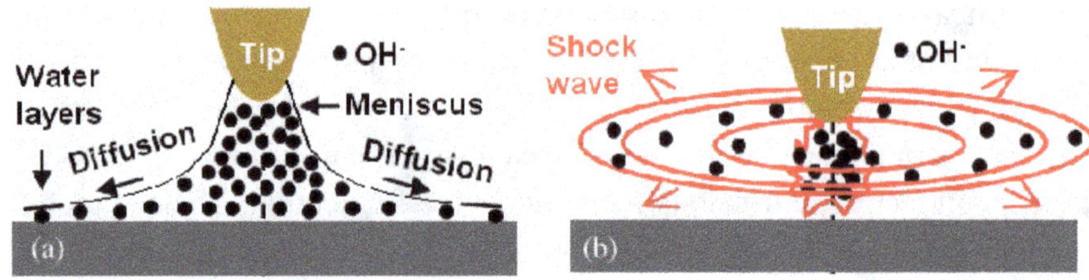

Fig. 2. (a) OH^- ionic diffusion model through the water meniscus and adsorbed water layers. (b) Spreading of OH^- by shock wave propagation. (*Adapted from Ref. 64.*)

highly localized in the vicinity of the tip apex. The larger base oxide is a result of oxidation at the later stage when the growth of core oxide saturates and the lateral diffusion of OH^- ions prevails. The authors proposed that OH^- diffusion through the water meniscus and adsorbed surface water layers is responsible for the formation of base oxides. This ionic diffusion mechanism, shown in Fig. 2(a), has been the only mode proposed for the spreading of OH^- reactants for the lateral growth of nano-oxide.[63]

Recently, a new mechanism for OH^- ionic spreading in AFM nano-oxidation has been proposed which involves the transport of OH^- reactants by transient shock wave propagation [see Fig. 2(b)].[64] Basically, under high humidity conditions, the biased tip can induce a nanoscale explosion through the electrical discharge of the air–water mixture condensed in the tip–surface nanojunction. Consequently, transient shock waves are generated and their propagation spreads the OH^- oxidants over a radial distance of several micrometers. The initiation of the nanoexplosion and associated shock wave is highly stochastic, and either one or two shock events can be observed in a single bias duration. Xie *et al.* reported that such a tip-induced shock wave propagation leads to the formation of low-aspect-ratio ultrathin nano-oxide disks.[64] When there is one shock event and the OH^- oxidants are spread out, a single disk oxide is formed [see Fig. 3(a)]. In the case of two shock events, a double-disk oxide is generated [see Fig. 3(b)]. In most cases, there is only one shock event, and the spreading of OH^- oxidants is not complete, so nano-oxide in the configuration of central-dot/outer-disk is observed, as shown in Fig. 3(c). The AFM images shown in Figs. 3(d)–3(f) illustrate the stochastic nature of the shock occurrence. Here the tip is scanned line by line over a Si surface area of 1×1 μm^2 to fabricate a square oxide. When there is no shock event, a well-defined square oxide is

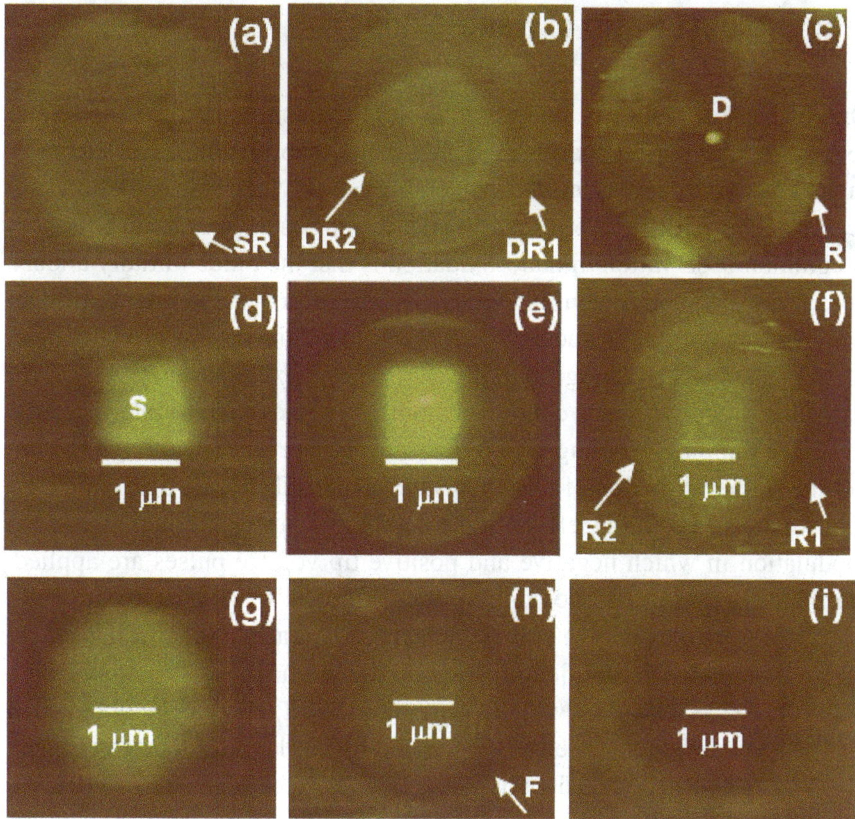

Fig. 3. (a–c) AFM images showing the formation of single-disk, double-disk and central-dot/outer-disk nano-oxide structures by nanoexplosion and shock wave propagation. (d–f) AFM images showing square oxide formation with zero (d), one (e) and two (f) shock events, respectively. AFM images showing disk oxide before (g), after partial (h), and prolonged (i) etching, respectively. Images (a–c) are adapted from Ref. 64.

obtained [see Fig. 3(d)]. In some cases, one or two shock events may be initiated stochastically by the bias during tip scanning, and thus structures with one or two disk oxides overlapping the square oxide are obtained, as shown in Fig. 3(e) and Fig. 3(f), respectively. The chemical identity of the disk is verified by dilute hydrofluoride (DHF) etching, and Figs. 3(g)–3(i) display the AFM images of the same disk recorded before, after partial, and prolonged DHF etching, respectively. The shock-wave-based disk oxide formation is useful for the fabrication of microscale ultrathin oxides for nonvolatile memory devices.

9.5. Aspect Ratio of Nano-oxide

One of the key issues in nano-oxidation is the aspect ratio of the oxide formed. A high aspect ratio is desirable in nanofabrication, as an increase in the aspect ratio represents a considerable gain in electronic depletion and scaling toward denser two-dimensional electron gas device architectures.[65] A common observation in nano-oxidation is that the oxide initially undergoes fast growth, and then its vertical growth saturates and lateral growth dominates if a prolonged negative tip bias is maintained. This growth mode is a result of space charge buildup[19–22] in the oxide formed. As nano-oxidation proceeds, space charges accumulate at the oxide/Si interface due to charge trapping, forming an interface energy barrier high enough to hinder the further diffusion of OH^- reactants to the substrate. To minimize or remove the space charge effect, some authors[20,65] used alternating current modulation in which negative and positive tip voltage pulses are applied alternatively during nano-oxidation. Since trapped space charges are neutralized during the positive voltage pulses, OH^- reactant diffusion through the growing oxide is significantly promoted in the negative bias pulses, and therefore oxide structures with an improved aspect ratio can be obtained.

Xie *et al.* demonstrated a new method for oxidation enhancement that is associated with the transient dynamic charge effect induced by switching the tip bias polarities during AFM nano-oxidation.[66] Four different voltage cycles (i)–(iv) were used for nano-oxidation (see Fig. 4). In cycle (i), the tip bias was ramped from $-12\,V$ to $12\,V$ and back to $-12\,V$ again, and it included both negative and positive bias polarity. The nano-oxide generated by cycle (i) exhibits a higher aspect ratio than that generated by cycle (iii), involving only negative polarity [see Fig. 4(a)]. The aspect ratio of the former was measured to be 13–18 times that of the latter. The authors observed upward cantilever deflection in the positive tip voltage region [see Figs. 4(c) and 4(d)], and on the basis of the downward deflection observed in Fig. 4(b) they attributed the upward deflection to the enhanced growth of nano-oxide under positive tip voltage. Usually, nano-oxidation proceeds only under a negative tip voltage as it is essentially an anodic reaction. Xie *et al.* proposed a charge pump effect model[66] to explain the seemingly contradictory observations. Their model is based on the space charge formation and neutralization in negative and positive bias phases, respectively. In the

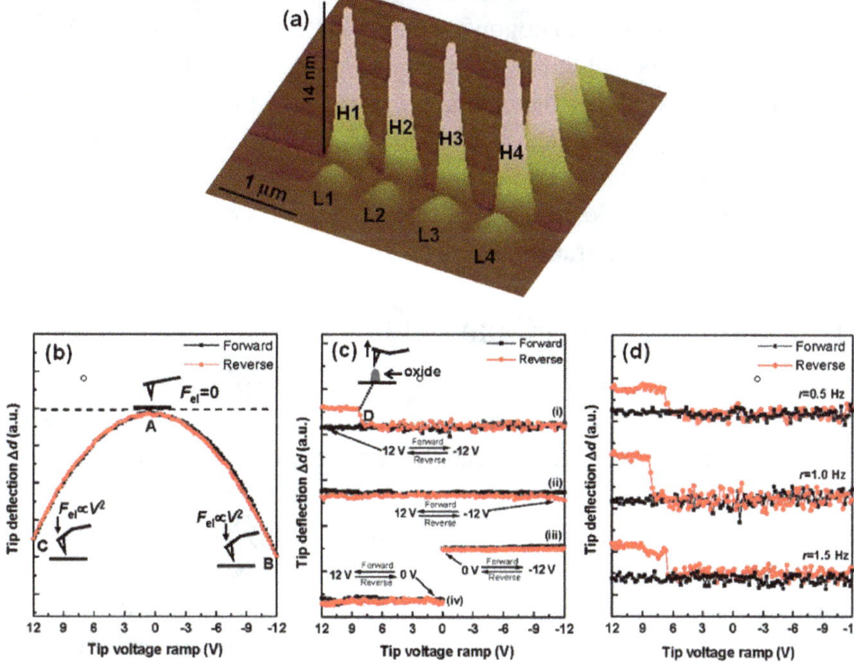

Fig. 4. (a) AFM image of high-aspect-ratio oxides H1–H4 and low-aspect-ratio oxides L1–L4 generated by voltage cycles (i) and (iii), respectively. (b) Control deflection curves obtained for cycle (i) under conditions of large tip–surface separation without oxide formation. Point A is the reference point where voltage $V = 0$ V, and deflection $d = 0$. Points B and C correspond to maximum cantilever downward bending when $V = -12$ and 12 V, respectively. (c) Cantilever deflection spectra collected during nano-oxidation using cycles (i)–(iv), respectively. (d) Deflection spectra collected when performing cycle (i) at rates of $r = 0.5$, 1.0, 1.5 Hz, respectively. (*Adapted from Ref. 66.*)

first phase, the usual anodic oxidation takes place, and at a certain stage the vertical growth of oxide saturates due to the buildup of space charges at the oxide/Si interface. The interface space charges create an energy barrier that hinders the diffusion of OH^- ions to Si for further oxidation reaction. If the negative tip bias is maintained, the lateral diffusion of OH^- oxidants will dominate, thus leading to the formation of low-aspect-ratio oxide, as observed in most cases. However, in cycle (i), the tip bias is quickly swept from negative to positive polarity, so the space charge is neutralized by electrons flowing into the Si substrate. The neutralization of space charges removes the energy barrier so that the OH^- ions, driven by the negative potential of the oxide, continue their diffusion and reaction. The switching of the two phases is analogous to the operation of a Dickson charge pump,[67] and is responsible for the enhanced oxide vertical growth under positive tip bias. Note that the charge pump model works only under certain voltage cycling frequencies, as it depends on the dynamical matching of polarity switching and space charge formation. These results reflect the complexity of nano-oxidation, and highlight the importance of transient charge dynamics in determining the oxide aspect ratio.

9.6. Media Used for Nano-oxidation

Most AFM nano-oxidation is carried out in ambient, with the formation of a water meniscus between the tip and the surface, and is thus dependent on the meniscus for the supply of OH^- oxidants [see Fig. 5(a)]. The size of the water meniscus depends strongly on the humidity conditions and the magnitude of the tip bias. Therefore, good control of the working environment and local electric field is necessary for the formation of oxide with the desired size and aspect ratio. Since the water meniscus is formed through the capillary and field-enhanced condensation of ambient water vapor, it is neutral in acidity, and can supply only species containing H and O atoms. This limits the applications of water-meniscus-based techniques, as many fabrication processes require the use of acidic solutions that contain etching species such as F^- and HF. To extend the capability of nano-oxidation, some authors used bulk solutions [see Fig. 5(b)].[68,69] For example, Kinser *et al.*[69] conducted nano-oxidation in an organic solution, and found that the inert organic solvent cannot suppress the anodic oxidation of Si, and

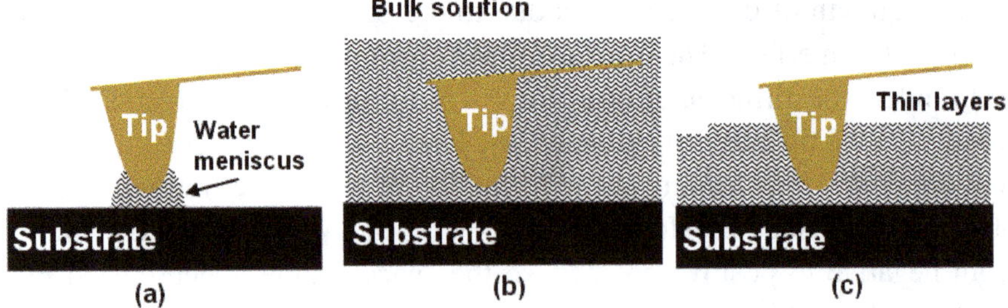

Fig. 5. Three media used in AFM nano-oxidation. (a) Water-meniscus-based medium. (b) Bulk-solution-based medium. (c) Thin-layer-based medium. (*Adapted from Ref. 70.*)

the finite amount of water dissolved in the nonpolar solvent can form a meniscus near the tip apex to facilitate oxide growth.

Xie *et al.*[70] presented an acidic thin-layer-based nano-oxidation approach that offers an intermediate working state between the local-meniscus-based and bulk-solution-based techniques [see Fig. 5(c)]. As shown in Fig. 5(c), the tip is operated in homogeneous thin layers with thicknesses on the submicrometer scale. This technique is different from the bulk-solution-based method in terms of the operational characteristics. Only the tip is in contact with the thin layers, while the cantilever is out of the liquid phase. Thus, there is a capillary force acting on the tip in this configuration, while such a capillary force would diminish significantly or completely in the bulk solution. This technique is also different from the local meniscus-based approach, because the former works in continuous thin layers while the latter relies on water condensation from ambient air. Moreover, the chemical content and acidity of the thin layers can be adjusted in Fig. 5(c), while it is difficult to vary the chemical composition of the air-condensed water meniscus in Fig. 5(a).

In their experiments, Xie *et al.* first prepared microscale droplets on a Si surface using DHF aqueous solution etching.[70] Such microdroplet formation is due to the coexistence of hydrophilic and hydrophobic sites on Si after etching. The authors then converted the microdroplets into acidic thin layers, shown in Fig. 5(c), by scanning an AFM probe repeatedly across the microdroplets. The thickness of the thin layers is controlled by adjusting the number of tip scanning cycles. The advantage of this method is that it allows the formation of both positive and negative patterns on Si. Mediated by the thin acidic layers, several reactions such as nano-oxidation and dissolution of nano-oxide and Si are possible due to the coexistence of OH^- oxidant and F^- etchant in the medium. Therefore, depending on the tip voltage magnitude and duration, one can obtain raised nano-oxide by activating the OH^--based anodic reaction. One can also generate depressed patterns by etching the nano-oxide and even the underlying Si substrate through the etching reactions $SiO_2 + 4H^+ + 6F^- \rightarrow SiF_6^{2-} + 2H_2O$ and $Si + 4h + 6F^- \rightarrow SiF_6^{2-}$.[70]

9.7. Physichemical Properties of Nano-oxide

Although AFM nano-oxidation has been studied for 20 years, most of the work is focused on the generation of oxide patterns, while knowledge of the chemical and dielectric properties of nano-oxide is very limited. In one early work, photoemission spectroscopy (PES) measurements were conducted to probe the chemical bonding states in nano-oxide.[71] The authors showed that AFM nano-oxidation produces chemically uniform, stoichiometric oxide in the form of SiO_2. Moreover, the chemical and structural properties of nano-oxide do not depend on the tip bias used in AFM nano-oxidation. The authors also studied the electrical property of nano-oxide on the basis of the electrostatic shift in binding energies. Since nano-oxidation is based on water dissociation, hydrogen may be incorporated, and thus H-related defects may be formed in the nano-oxide. However, PES cannot detect the presence of H atoms introduced during nano-oxidation in ambient AFM operation. This issue must be addressed, because H-bearing species such as Si–H and Si–OH have significant impact on the structure and dielectric performance of oxides.

Xie *et al.* performed time-of-flight secondary ion mass spectrometry (TOF-SIMS) and conducting AFM (cAFM) experiments to characterize AFM nano-oxide.[33] TOF-SIMS ultrashallow time profiling is very sensitive to monolayer surface chemistry, while cAFM is capable of measuring the local current–voltage (*IV*) curves of oxide. In their TOF-SIMS experiments, Xie *et al.* detected the presence of Si–H and Si–OH defects in the nano-oxide. These H-containing sites are formed by breaking the Si–O–Si network, and cause the nano-oxide to deviate from the ideal tetrahedral SiO_2 structure. In other words, the presence of Si–H and Si–OH sites partially terminates the silica network and causes the structure to be more porous and open. Xie *et al.* also observed higher leakage current and dielectric breakdown probability for the nano-oxide.[33] Such dielectric behavior is correlated with the formation of Si–H and Si–OH traps in nano-oxide. Electrons can be trapped and released from the trap states with minimum structure reconstruction and energy consumption. Once the density of Si–H and Si–OH traps reaches a critical value, complete conduction paths are formed, thus leading to oxide dielectric breakdown. The Si–H and Si–OH intrinsic traps in AFM nano-oxide induce the formation of extrinsic traps under electrical stress.

In another work, Xie and Chung *et al.* reported the chemical etching of ultrathin nano-oxide.[72] They used an AFM localized depth analysis to monitor the atomic layer-by-layer etching of nano-oxide. Insights into the growth mode and etching mechanism of nano-oxide were acquired on the basis of their etching results. It was proposed that the layer-by-layer oxide growth is enhanced by Si out-diffusion. The etching rate of ultrathin nano-oxide is dependent on the Si–OH silanol reactive sites. Thermal annealing reduces the content of silanol groups and enhances the chemical stability of AFM nano-oxide against etching.

9.8. Applications of Nano-oxidation

Apart from the fundamental studies on the mechanism and control of the aspect ratio in nano-oxidation, there are numerous works on the application of nano-oxidation in nanotechnology. There are two general applications of AFM nano-oxidation: (i) as an intermediate stage for further fabrication; (ii) as an integral part of the final devices. In the first case, the nano-oxide is often used as a mask for selective etching or a template for further assembly and growth. For example, Snow *et al.* fabricated nano-oxide on a Si substrate, and the oxide was then used as a mask for dry etching in a Cl_2 plasma generated by an electron cyclotron resonance source.[73] The combination of nano-oxidation with dry etching provides high selectivity, excellent linewidth control, vertical profile and smooth surface morphology. The authors' AFM nano-oxide was observed to withstand a 70 nm deep etch. Features with ~10 nm width were etched 30 nm deep by optimized conditions with minimum oxide removal.

Sugimura *et al.* fabricated pits for the selective plating of Au by etching AFM nano-oxide.[55] They first created a nano-oxide on Si covered with an ODS self-assembled monolayer (SAM). The nano-oxide was then preferentially etched by HF solution, such that depressed areas 5 nm lower than the surrounding ODS SAM surface were generated. The sample was then treated in an electroplating bath, and Au was selectively deposited on the etched pits while the ODS surface remained free of deposits.

Lin *et al.* demonstrated the catalytic function of nano-oxide in the growth of carbon nanotubes (CNTs).[45] They produced nickel oxide on a Ni surface using AFM nano-oxidation. After the removal of unoxidized Ni, nickel oxide nanodots were formed with a minimum size of ~30 nm. Vertically aligned bunched or single CNTs with diameters of 30–80 nm were selectively grown only on the oxidized surface regions, due to the catalytic role of the nano-oxide. The authors also observed that the tube diameter could be effectively controlled by tuning the size of the nickel oxide dots.

Campbell *et al.* demonstrated the use of AFM nano-oxide in the fabrication of nanoscale side-gated silicon field effect transistors (FETs).[23] Both the source-drain channels and the side gates were patterned by AFM nano-oxidation. The resulting oxide pattern was transferred onto the Si layer on an insulating SiO_2 layer for isolation. Devices with critical sizes as small as 30 nm were fabricated which exhibited current–voltage characteristics typical of transistors.

9.9. Concluding Remarks

We have reviewed both the fundamental and the technical aspects of AFM nano-oxidation. The growth models, OH^- oxidant spreading modes, and the chemical and dielectric properties of nano-oxide have been discussed. In addition, the technical development including the materials and media used for nano-oxidation, and the applications of nano-oxide in nanopatterning and nanodevice operation, have been summarized.

Acknowledgments

This work is supported by the NUS Nanoscience and Nanotechnology Initiative (NUSNNI), National University of Singapore.

References

1. G. Binnig, C. F. Quate and C. Gerber, *Phys. Rev. Lett.* **56**, 930 (1986).
2. G. Binnig, H. Rohrer, C. Gerber and E. Weibel, *Appl. Phys. Lett.* **40**, 178 (1982).
3. X. N. Xie, H. J. Chung, C. H. Sow and A. T. S. Wee, *Mater. Sci. Eng. R* **54**, 1 (2006).
4. D. S. Ginger, H. Zhang and C. A. Mirkin, *Angew. Chem. Int. Ed.* **43**, 30 (2004).

5. R. Garcia, R. V. Martinez and J. Martinez, *Chem. Soc. Rev.* **35**, 29 (2006).
6. S. Kramer, R. R. Fuierer and C. B. Gorman, *Chem. Rev.* **103**, 4367 (2003).
7. D. Wouters and U. S. Schubert, *Angew. Chem. Int. Ed.* **43**, 2480 (2004).
8. A. A. Tseng, A. Notargiacomo and T. P. Chen, *J. Vac. Sci. Technol.* **B23**, 877 (2005).
9. D. B. Gates, Q. B. Xu, M. Stewart, D. Ryan, C. G. Willson and G. M. Whitesides, *Chem. Rev.* **105**, 1171 (2005).
10. M. Geissler and Y. Xia, *Adv. Mater.* **16**, 1249 (2004).
11. D. Fotiadis, S. Scheuring, S. A. Muller, A. Engel and D. J. Muller, *Micron* **33**, 385 (2002).
12. X. N. Xie, H. J. Chung and A. T. S. Wee, in *COSMOS*, ed. A. T. S. Wee, Vol. 3, No. 1, pp. 1–21 (World Scientific, 2007).
13. A. E. Gordon, R. T. Fayfield, D. D. Litfin and T. K. Higman, *J. Vac. Sci. Technol. B* **13**, 2805 (1995).
14. T. Teuschler, K. Mahr, S. Miyazaki, M. Hundhausen and L. Ley, *Appl. Phys. Lett.* **67**, 3144 (1995).
15. H. Sugimura, N. Kitamura and M. Mushuhara, *Jpn. J. Appl. Phys.* **33**, L143 (1994).
16. M. Yasutake, Y. Y. Ejiri and T. Hattori, *Jpn. J. Appl. Phys.* **32**, L1021 (1993).
17. T. Teuschler, K. Mahr, S. Miyazaki, M. Hundhausen and L. Ley, *Appl. Phys. Lett.* **66**, 2499 (1995).
18. D. Stíevenard, P. A. Fontaine and E. Dubois, *Appl. Phys. Lett.* **70**, 3272 (1997).
19. J. A. Dagata, T. Inoue, J. Itoh and H. Yokoyama, *Appl. Phys. Lett.* **73**, 271 (1997).
20. J. A. Dagata, T. Inoue, J. Itoh, K. Matsumoto and H. Yokoyama, *J. Appl. Phys.* **84**, 6891 (1998).
21. J. A. Dagata, F. Perez-Murano, G. Abadal, K. Morimoto, T. Inoue, J. Itoh and H. Yokoyama, *Appl. Phys. Lett.* **76**, 2710 (2000).
22. J. A. Dagata, F. Perez-Murano, C. Martin, H. Kuramochi and H. Yokoyama, *J. Appl. Phys.* **96**, 2386 (2004).
23. P. M. Campbell, E. S. Snow and P. J. McMarr, *Appl. Phys. Lett.* **66**, 1388 (1995).
24. S. C. Minne, H. T. Soh, Ph. Flueckiger and C. F. Quate, *Appl. Phys. Lett.* **66**, 703 (1995).
25. S. C. Minne, Ph. Flueckiger, H. T. Soh and C. F. Quate, *J. Vac. Sci. Technol. B* **13**, 1380 (1995).
26. S. C. Minne, S. R. Manalis, A. Atalar and C. F. Quate, *J. Vac. Sci. Technol. B* **14**, 2456 (1996).
27. S. C. Minne, J. D. Adams, G. Yaralioglu, S. R. Manalis, A. Atalar and C. F. Quate, *Appl. Phys. Lett.* **73**, 1742 (1998).
28. A. Dorn, M. Sigrist, A. Fuhrer, T. Ihn, T. Heinzel, K. Ensslin, W. Wegscheider and M. Bichler, *Appl. Phys. Lett.* **80**, 252 (2002).
29. M. Tachiki, H. Seo, T. Banno, Y. Sumikawa, H. Umezawa and H. Kawarada, *Appl. Phys. Lett.* **81**, 2854 (2002).
30. X. N. Xie, H. J. Chung, C. H. Sow and A. T. S. Wee, *Appl. Phys. Lett.* **84**, 4914 (2004).
31. X. N. Xie, H. J. Chung, H. Xu, X. Xu, C. H. Sow and A. T. S. Wee, *J. Am. Chem. Soc.* **126**, 7665 (2004).

32. X. N. Xie, H. J. Chung, C. H. Sow and A. T. S. Wee, *Appl. Phys. Lett.* **86**, 023112 (2005).
33. X. N. Xie, H. J. Chung, C. H. Sow and A. T. S. Wee, *Appl. Phys. Lett.* **86**, 192904 (2005).
34. F. S.-S. Chien, J.-W. Chang, S.-W. Lin, Y.-C. Chou, T. T. Chen, S. Gwo, T.-S. Chao and W.-F. Hsieh, *Appl. Phys. Lett.* **76**, 360 (2000).
35. F. S.-S. Chien, Y. C. Chou, T. T. Chen, W.-F. Hsieh, T.-S. Chao and S. Gwo, *J. Appl. Phys.* **89**, 2465 (2001).
36. R. Held, T. Vancura, T. Heinzel, K. Ensslin, M. Holland and W. Wegscheider, *Appl. Phys. Lett.* **73**, 262 (1998).
37. E. S. Snow and P. M. Campbell, *Science* **270**, 1639 (1995).
38. T. Schmidt, R. Martel, R. L. Sandstrom and Ph. Avouris, *Appl. Phys. Lett.* **73**, 2173 (1998).
39. K. Matsumoto, Y. Gotoh, T. Maeda, J. A. Dagata and J. S. Harris, *Appl. Phys. Lett.* **76**, 239 (2000).
40. E. S. Snow, D. Park and P. M. Campbell, *Appl. Phys. Lett.* **69**, 269 (1996).
41. D. Wang, L. Tsau, K. L. Wang and P. Chow, *Appl. Phys. Lett.* **67**, 1295 (1995).
42. E. S. Snow, P. M. Campbell, R. W. Rendell, F. A. Buot, D. Park, C. R. K. Marrian and R. Magno, *Appl. Phys. Lett.* **72**, 3071 (1998).
43. V. Bouchiat, M. Faucher, C. Thirion, W. Wensdorfer, T. Fournier and B. Pannertier, *Appl. Phys. Lett.* **79**, 123 (2001).
44. J.-H. Hsu, H.-W. Lai, H.-N. Lin, C.-C. Chuang and J.-H. Huang, *J. Vac. Sci. Technol. B* **21**, 2599 (2003).
45. H.-N. Lin, Y.-H. Chang, J.-H. Yen, J.-H. Hsu, I.-C. Leu and M.-H. Hon, *Chem. Phys. Lett.* **399**, 422 (2004).
46. Y. Kim, J. Zhao and K. Uosaki, *J. Appl. Phys.* **94**, 7733 (2003).
47. M. Rolandi, C. F. Quate and H. Dai, *Adv. Mater.* **14**, 191 (2002).
48. N. Farkas, J. C. Tokash, G. Zhang, E. A. Evans, R. D. Ramsier and J. A. Dagata, *J. Vac. Sci. Technol. A* **22**, 1879 (2004).
49. N. Farkas, G. Zhang, E. A. Evans, R. D. Ramsier and J. A. Dagata, *J. Vac. Sci. Technol. A* **21**, 1188 (2003).
50. Y. Takemura, S. Hayashi, F. Okazaki, T. Yamada and J. Shirakashi, *Jpn. J. Appl. Phys.* **44**, L285 (2005).
51. X. N. Xie, H. J. Chung, C. H. Sow and A. T. S. Wee, *Chem. Phys. Lett.* **388**, 446 (2004).
52. H. Sugimura, K. Okiguchi, N. Nakagiri and M. Miyashita, *J. Vac. Sci. Technol. B* **14**, 4140 (1996).
53. J. Kim, Y. Oh, H. Lee, Y. Shin and S. Park, *Jpn. J. Appl. Phys.* **37**, 7148 (1998).
54. D. C. Tully, K. Wilder, J. M. J. Fréchet, A. R. Trimble and C. F. Quate, *Adv. Mater.* **11**, 314 (1999).
55. H. Sugimura, O. Takai and N. Nakagiri, *J. Electroanal. Chem.* **473**, 230 (1999).

56. H. Sugimura, T. Hanji, K. Hayashi and O. Takai, *Adv. Mater.* **14**, 524 (2002).
57. W. Lee, E. R. Kim and H. Lee, *Langmuir* **18**, 8375 (2002).
58. M. Rolandi, I. Suez, H. Dai and J. M. J. Fréchet, *Nano Lett.* **4**, 889 (2004).
59. R. V. Martinez and R. Garcia, *Nano Lett.* **5**, 1161 (2005).
60. M. Tello, R. Garcia, J. A. Martin-Gago, N. F. Martinez, M. S. Martin-Gonzale, L. Aballe, A. Baranov and L. Gregoratti, *Adv. Mater.* **17**, 1480 (2005).
61. H. Kuramochi, K. Ando and H. Yokoyama, *Surf. Sci.* **542**, 56 (2003).
62. H. Kuramochi, F. Pérez-Murano, J. Dagata and H. Yokoyama, *Nanotechnology* **15**, 297 (2004).
63. T.-H. Fang, *Microelectron. J.* **35**, 701 (2004).
64. X. N. Xie, H. J. Chung, Z. J. Liu, S. W. Yang, C. H. Sow and A. T. S. Wee, *Adv. Mater.* **19**, 2618 (2007).
65. D. Graf, M. Frommenwiler, P. Studerus, T. Ihn, K. Ensslin, D. C. Driscoll and A. C. Gossard, *J. Appl. Phys.* **99**, 053707, 2006.
66. X. N. Xie, H. J. Chung, C. H. Sow and A. T. S. Wee, *Appl. Phys. Lett.* **91**, 243101 (2007).
67. J. Dickson, *IEEE J. Solid-State Circuits* **11**, 374 (1976).
68. I. Suez, S. A. Backer and J. M. J. Fréchet, *Nano Lett.* **5**, 321 (2005).
69. C. R. Kinser, M. J. Schmitz and M. C. Hersam, *Nano Lett.* **5**, 91 (2005).
70. X. N. Xie, H. J. Chung, C. H. Sow and A. T. S. Wee, *Adv. Funct. Mater.* **17**, 919 (2007).
71. M. Lazzarino, S. Heun, B. Ressel, K. C. Prince, P. Pingue and C. Ascoli, *Appl. Phys. Lett.* **81**, 2842 (2002).
72. X. N. Xie, H. J. Chung, C. H. Sow and A. T. S. Wee, *J. Appl. Phys.* **99**, 044301 (2006).
73. E. S. Snow, W. H. Juan, S. W. Pang and P. M. Campbell, *Appl. Phys. Lett.* **66**, 1729 (1995).

Xian Ning Xie obtained his PhD in Chemistry from the National University of Singapore. He joined NUSNNI in 2002, and works as Senior Research Fellow. Dr Xie's research interest is in AFM nanolithography for localized and site-specific pattern formation. Specifically, the nano-oxidation of semiconductors and bias-assisted patterning of polymers are the two focuses of his work. Recently, his work has been extended to the search for solutions to environment and energy problems.

Dr. Chung obtained his PhD degree from the National University of Singapore. He joined Materials Science and Technology of Polymers (MTP) chaired by Professor G. Julius Vancso at The University of Twente, The Netherlands in 2008 as a Postdoctoral Fellow. He then started to work as a Research Engineer at the Institute of Materials Research and Engineering (IMRE) in Singapore. He is now a research staff in Prof. Morita's Laboratory in Osaka University, Japan under the international fellowship funding sponsored by Agency for Science, Technology and Research (A*STAR) (Singapore). Dr. Chung's research interest is in advanced AFM applications and characterization techniques. Recently, his work has extended to site-specific manipulation, patterning, and force spectroscopy at single-atom level by using UHV-AFM at both room temperature and low temperature.

Andrew T. S. Wee is Professor of Physics and currently Dean of the Faculty of Science at the National University of Singapore. He graduated with a BA (Hons) in Physics (1984) from the University of Cambridge, and a DPhil. (1990) from the University of Oxford on a Rhodes Scholarship. Prof. Wee is also Director of the Surface Science Laboratory. He has published over 300 papers in internationally refereed journals in the field of surface and nanoscale science, and is an editor of several academic journals, including *Applied Physics Letters*, *Surface and Interface Analysis*, *International Journal of Nanoscience* and *Surface Review and Letters*.

Nanolithography of Organic Films Using Scanning Probe Microscopy

Chapter

10

Jegadesan Subbiah*, Sajini Vadukumpully*, and Suresh Valiyaveettil*,†

Abstract

Patterning of polymeric materials in nanoscale dimensions has attracted attention because of its perceived applications in the fabrication of data storage devices, optoelectronics, displays and nanosensors. Recently, the scanning probe microscope (SPM) has been used to manipulate and pattern soft materials down to atomic resolution. This chapter highlights the similarities and differences between the various AFM-based nanolithography techniques for nanopatterning using electrostatic and electrochemical methods, and provides a brief overview of the challenges and future developments. The principles and specific approaches underlying each technique for nanopattern formation are discussed in detail.

10.1. Introduction

Micro/nano-fabrication techniques in the semiconductor industry depend largely on the lithographic process to make microelectronic circuits and devices.[1,2] All electronic devices are developed on semiconducting chips built from lithographic processes.[3–7] In conventional lithographic

*Department of Chemistry, National University of Singapore, 3 Science Drive 3, Singapore 117543.
†Email: chmsv@nus.edu.sg

techniques, the first step involves the formation of a pattern on the resist layer on top of a substrate by selectively irradiating it with electrons, ions or photons, followed by chemical etching.[8,9] These indirect patterning approaches compromise the chemical purity of the structure generated and limit the type and number of materials that can be patterned. Most of these conventional lithographic methods either will break down or are poorly controlled when the feature sizes go down to the nanometer scale.[10] These drawbacks lead to the development of new lithographic methods such as direct writing, where materials can be patterned directly on the substrate at the nanoscale. Development of new lithographic techniques is the key to the fabrication of nanoscale structures using organic and inorganic materials. Nanofabrication can be achieved both by the extension of current micro-fabrication techniques into a smaller size regime and by the development of new techniques or tools such as the scanning probe microscope (SPM).[11] Many lithographic methods, such as UV lithography,[3–7] electron beam lithography[8] and imprint lithography,[12] have been developed to fabricate nanoscale structures from inorganic and organic films. However, the drive to miniaturize devices without compromising on the performance led to the development of new fabrication methods at the nanoscale. Among the various techniques, the SPM-based lithographic technique is a versatile tool for manipulating as well as for visualizing the surface at the nanoscale. Due to the successful application of organic materials in electronic devices such as LEDs and display devices, there is a growing interest and need for the development of nanoscale structures using organic materials. However, there are only a few reviews on the fabrication of nanoscale structures with organic/polymeric materials using AFM lithography.[11,13–19] The main merit of the AFM-based lithographic technique is that the machined scale of the structure is entirely dependent on the geometry of the AFM probe. Hence, it is essential to explore various polymeric and organic materials for the fabrication of functional nanoscale structures for nanodevices. In addition, understanding the mechanism of the nanofabrication process is crucial to select suitable materials for potential applications.

This chapter focuses on the various AFM lithographic techniques for patterning organic surfaces, including polymers and oligomers, at the nanoscale. Here, we take advantage of the nanomanipulation ability of the

AFM probe to fabricate nanostructures on various polymer films together with the probe biasing or mechanical force.

10.1.1. *Principles of AFM lithography*

AFM lithography is a promising method for the fabrication of patterns with high resolution. The sharpness of the tip and accurate positioning allow the AFM to make a pattern on the nanometer scale.[20–23] The working principle of AFM nanolithography is based on the interaction between the AFM tip and the substrate.

There are various interactions experienced by an AFM probe when moving on a surface. Understanding these interactions is important for obtaining the exact information of a surface. Figure 1 shows the typical variation in the interaction potential between the probe and the sample as the tip approaches the sample. When the separation between the tip and the sample is large, the attractive force is small. This attractive force will be increased when the separation is less than hundreds of angstroms. AFM utilizes the force–distance relationship to provide the surface properties of the sample when scanning the sample. The forces essentially experienced by the tip are (i) van der Waals, (ii) surface tension, (iii) electrostatic force and (iv) chemical force.

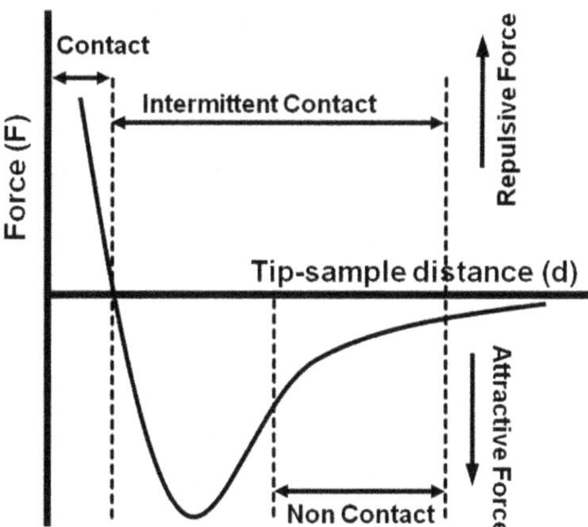

Fig. 1. Distance dependence of van der Waals and electrostatic forces compared to the typical tip–surface separations in the contact mode (CM), noncontact mode (NCM) and intermittent contact mode.

At large distances and in the absence of a meniscus force, the interaction is dominated by van der Waals forces, which increase in magnitude when the probe approaches the sample surface. Electrostatic forces are dominant when there is a voltage difference between the tip and the sample; an attractive Coulomb force pulls them together. Chemical forces result from the interactions of the electrons at the apex of the tip with those at the sample surface. They are the same forces that are responsible for covalent or hydrogen bonds and cause repulsion, which is measured in contact AFM.

The AFM probe can induce various patterns on soft films deposited on the substrate when suitable force such as a mechanical or electrical field is applied between the probe and the substrate. Based on this probe-induced principle, AFM-based lithography can be classified into two categories, according to the probe operation on the substrate: (i) mechanical probe AFM nanolithography and (ii) biased probe AFM nanolithography.

Various nanolithographic methods have been developed using the above principles. Representative methods are nanoindentation,[24] nanografting,[25,26] nanomanipulation,[27] dip pen nanolithography,[28] anodic oxidation,[29] electrochemical deposition,[30] field-induced deposition[31] and thermomechanical writing.[32] The main advantage of the AFM-probe-based nanolithographic techniques is that the machined scale of the structure is entirely determined by the geometry of the AFM probe. This chapter elaborates on some of the above-mentioned techniques used in the current research for future applications.

10.1.2. *Mechanical probe nanolithography*

Since the invention of scanning probe methods,[11,13–16] considerable efforts have been made to manipulate and structure matter at the nano and molecular levels. For manipulating matter at the nanoscale, mechanical probe methods are very convenient. Due to the small size of the probes, which interacts directly with surfaces, a resolution of a few nanometers should be achievable. Mechanical-probe-based direct structuring of organic thin films has been performed by scratching or indentation experiments.[33–38] Local topographical features can also be created by thermomechanical writing using a heated AFM tip for local melting of a polymer substrate.[39–49] The localized deposition of a foreign substance on a surface has been demonstrated with a method called "dip pen" nanolithography (DPN).[50]

10.1.2.1. *Nanofabrication using self-assembled monolayers*

Self assembled monolayers (SAMs) have attracted tremendous attention in nanofabrication because of their utility as resists for pattern transfer and as templates for directing the direction and growth of metals, polymer films and protein layers.[51,52] They offer a flexible and convenient way to give surface functionality in a chemically defined way. This technique provides relatively ordered structures at the molecular scale, with interesting functional groups at the surface. To exploit this chemical functionality at the nanometer scale, it is attractive to take advantage of the nanomanipulating ability of the SPM to create functional nanoscale structures. The nanopatterned SAMs produced to date via scanning probe lithography (SPL) have primarily been negative patterns, where the surfaces of the patterned regions are lower in height than the surrounding matrix areas. The development of nanoscale structures on SAMs can be divided into three categories: elimination lithography,[53–61] addition lithography and substitution lithography.[39,62–72]

In elimination lithography, the patterns are produced by selective removal of surface atoms or adsorbate molecules within the adsorbed layer. Addition lithography involves molecular materials deposited on the substrate directly from a tip. A probe coated with molecular ink is brought into contact with a "bare" substrate. The ink gets transferred from the probe to the surface, and this process is known as DPN. Substitution lithography denotes *in situ* mechanical or electrochemical pattern fabrication strategies. Substitution approaches are subdivided into two categories: (i) initial elimination of the SAM film, followed by (ii) *in situ* addition of another component to the exposed substrate.

Various AFM-assisted lithographic techniques have been developed using SAMs. In this lithography process, the SAM is first imaged with low contact force, and then fabrication methods are chosen. The patterns are then created on the surface by either depositing or removing the materials on the surface using the AFM probe.

10.1.2.2. *Scanning probe anodization*

SPM-based local anodic oxidation is used for the nanoscale patterning of metals, semiconductors and organic thin films. In this lithographic process, an anodically oxidizable film is used as a resist to be patterned by scanning probe anodization.

The mechanism of scanning probe anodization is as follows. The tip–sample junction of an SPM is connected through the water column created by the capillary of the adsorbed water. This junction acts as a small electrochemical cell. By applying a small positive potential, the tip will act as anode and the sample as cathode. It induces anodic oxidation on the surface of the sample beneath the sample tip (Fig. 2).[70]

10.1.2.3. *Thermomechanical writing*

The surface modification capability of scanning probes such as mechanical scratch and manipulation of molecules and nanotubes have motivated efforts to create data storage devices based on scanning probe techniques for storing data at much higher densities. The method known as thermomechanical writing is based on the indentation of an AFM tip on a polymer surface, which is softened locally by the combined effect of heat and mechanical force from the AFM tip. This technique was developed by the IBM Zurich research group,[44,73–75] who achieved a storage density of up to 30 Gb/in^2.

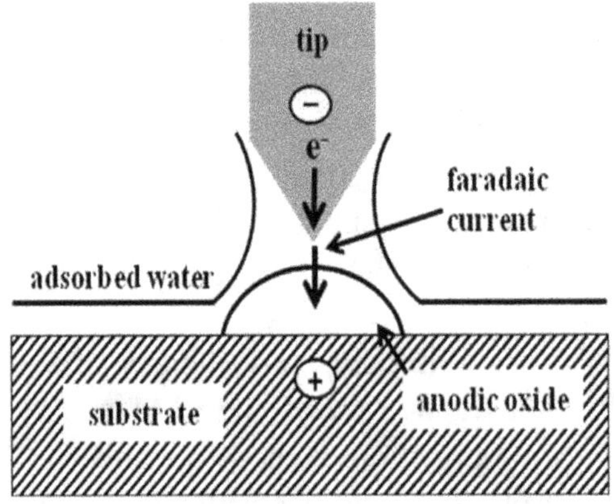

Fig. 2. Schematic illustration of scanning probe anodization. (*Adapted from Ref. 70.*)

Here, the AFM typically operates on the microsecond timescale, whereas the conventional magnetic storage operates on the nanosecond timescale. To improve the data storage speed, MEMS-based arrays of cantilever operating in parallel were used.[45–49] In this technique, high data transfer rates are achievable through the massive parallel operations of such tips.

The contact between the probe arrays and the polymer sample is realized by adjusting the z-piezo voltage, and only one feedback is used to control all the probes. Additional approaching sensors are integrated into the corners of the chip to control the approach to the storage medium. These sensors can provide feedback signals to adjust the z-actuator until uniform contact with the medium is established. During data writing, the chip is scanned over an area called the "storage field," and each cantilever of the array writes and reads data only in its own storage field. For parallel operation on a large scale, a microscanner is developed to control the position of the probe array and the sample.[45] A computer-controlled write/read scheme addresses the 32 cantilevers of one row in parallel. Writing is performed by connecting the row for 20 μs to a high, negative voltage and simultaneously applying data inputs to the 32 column lines. The row-enabling and column-addressing scheme supplies a heater current to all cantilevers. Selected cantilevers heated to high temperatures generate an indentation, while those with low temperatures make no indentation.

10.1.2.4. *Dip pen nanolithography*

DPN uses an AFM tip to deliver small molecules to a surface through a solvent meniscus, which is naturally formed under ambient conditions. In this technique, an AFM tip is coated with a thin film of thiol derivative (or ink) which is to be transferred. When the tip is placed close to the surface at high humidity, a tiny drop of water condenses between the tip and the surface, and the thiol molecules migrate from the tip to the substrate through the water meniscus. The tip is then moved across the surface to deliver the molecules in a predetermined pattern.[66,76–91] The chemisorption of ink molecules to the substrate is the driving force for the transportation of the ink from the AFM tip through the water to the substrate. Due to the high reactivity of thiol functional groups with gold, a compact monolayer can be formed as the ink molecules migrate onto the substrate.

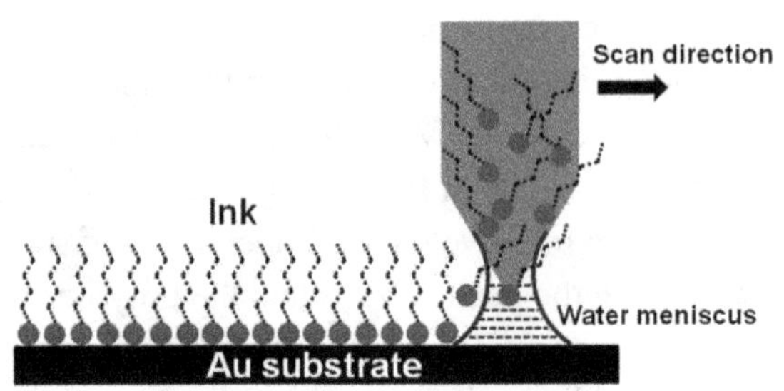

Fig. 3. Schematic representation of DPN. (Adapted from Ref. 66.)

The success of DPN depends on two main factors: a narrow spatial deposition of patterning molecules, and self-assembly of the functionalized ink molecules to form a compact monolayer.[92] The transportation rate and the linewidth of the pattern depend on the size of the water meniscus, which can be controlled by adjusting the relative humidity. The tip–substrate contact time and scan speed can also influence DPN resolution. Faster scan speeds can produce narrower lines. However, very high speed results in failure of effective transportation of the ink molecules.[93]

For example Piner *et al.* prepared a dot pattern by DPN to demonstrate the diffusion properties of the ink molecules. The tip was coated with 1-octadecanethiol or 16-mercaptohexadecanoic acid and brought into contact with the gold substrate for a set period of time. The size of the dot pattern depends on the contact time and the diffusion properties of ink molecules on the substrate.[50]

DPN is capable of creating structures made of various materials, such as metal, inorganic compounds, organic molecules and biological species.[94–96] The fabrication of a wide range of functional structures by DPN has been demonstrated and some of the typical structures are high resolution metallic structures, magnetic patterns, polymer brush arrays and biological devices. Ali *et al.* demonstrated the direct transfer of Au nanoparticles from the tip to a silica substrate.[97] The direct deposition of Au and Pd nanocrystals on a mica substrate using hydrosol inks was also described.[79] The deposition of Au nanoparticles on SAM-covered Au surface through S–Au linkage was presented by Garno *et al.*[65] The conversion of soluble Au (III) ion to insoluble Au(0) metal was attributed to surface-induced reduction of noble metal ions during the adsorption process. Ivanisevic and Mirkin used hexamethyldisilazane (HMDS) as ink to pattern organic materials on Si/SiO_x and oxidized GaAs semiconducting substrates.[83] In addition to the inorganic and organic nanostructures, biological molecules can also be patterned using the DPN process. Antirabbit IgG (immunoglobulin)[77] or active protein arrays are fabricated using the process. In addition, the process has been extended to the delivery of many organic materials such as thiols, silazanes, alkynes, polymers and proteins.[77–79,82–87]

10.1.3. *Biased probe nanolithography*

The use of an AFM tip in nanopatterning and nanomanipulation of surface can be extended by employing an electrically biased probe to explore many physicochemical phenomena on the organic substrate. A biased AFM tip can create a highly localized electric field of 10^8–10^{10} V/m between the substrate and the tip. Depending on the substrate material, this localized high electric field can initiate various processes, such as electrostatic, electrochemical and dielectric breakdown of polymers, to facilitate pattern formation at the nanoscale. This strategy was applied to develop nanopatterning processes such as electrostatic nanolithography and electrochemical nanolithography using organic and polymeric materials. In electrostatic nanolithography, the applied bias induces the nanopattern on polymer film through the combined process of localized Joule heating and electrohydrodynamic (EHD) instability of the film. Here, physical changes on the film create the nanoscale pattern. But, in electrochemical nanolithography, the patterns are formed on the polymer film through an electrochemical process, in which the polymer surface is chemically modified to create the functional nanostructure on the film. In this process, a suitable polymer, such as an electroactive polymer, is used to initiate the electrochemical process using a water meniscus as electrolyte under the high electric field.

10.1.3.1. *Electrostatic nanolithography*

Electrostatic nanolithography is an SPM-based lithographic technique in which patterns are drawn on thin polymer/organic films on a solid substrate using a biased AFM tip.[98] The electric field has proven to be a suitable tool for contactless, rapid and parallel manipulation of small particles by using electrophoretic forces such as in electrodeposition. The polymer/organic film can be modified directly at the nanoscale with the combined action of the AFM and electric fields with nanoscale manipulation capability. In this method, nanopatterns are created on the substrate due to Joule heating of the polymer film without any chemical modification when the electric field is applied between the tip and the film.[98]

(a) Nanopatterning of poly(methacrylic acid) (PMAA) films

The controlled patterning ability in thin film of PMAA on Si(100) substrate using electrostatic nanolithography is studied with the optimization of writing parameters such as writing speed and bias voltages. Figure 4 shows the nanolines drawn on a PMAA-coated Si(100) substrate at different voltages varying from −4 to −11 V at an AFM tip speed of 0.5 μm/s. The observed linewidths are 62, 78, 113, 135 and 172 nm, respectively. The first two lines drawn at −4 and −5 V were raised patterns, and the last two lines, drawn at −9 and −11 V, were grooved patterns. At −7 V, both raised (width 65 nm) and grooved patterns or nanopits (width 48 nm) were formed.

The grooved patterns are formed at a higher voltage (−11 V) with an increase in applied bias time. Figure 4(b) shows the nanopits of uniform size,

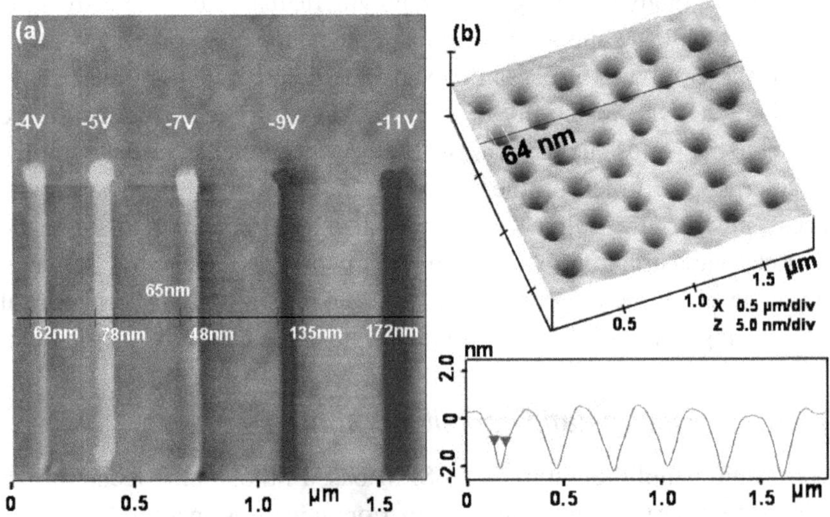

Fig. 4. (a) AFM image of the nanolines drawn with a tip speed of 0.5 μm/s at different applied voltages. (b) Array of nanopits formed at −12 V with a diameter of 64 nm and the corresponding height profile of the pattern. (*Adapted from Ref. 99.*)

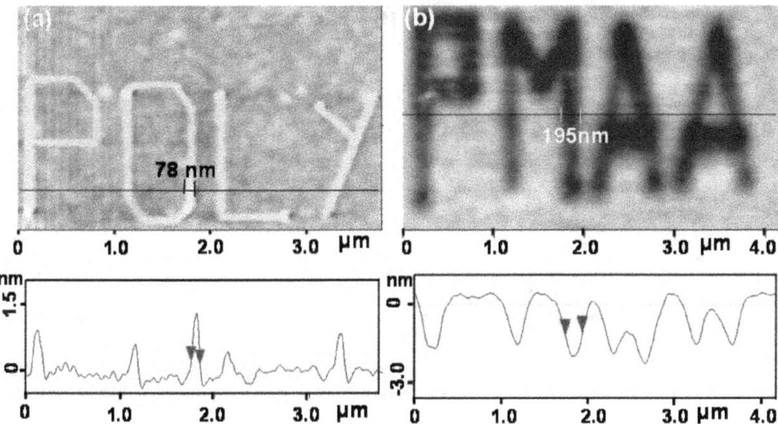

Fig. 5. AFM images (top) and height profiles (bottom) of (a) raised lines of letters, 78 nm wide, formed at -5 V with a tip speed of 0.5 μ ms^{-1}; and (b) grooved patterns, 195 nm wide, drawn at a tip voltage and speed of -12 V and 0.5 μm/s, respectively (*Adapted from Ref. 99.*)

64 nm in diameter, fabricated at a voltage of -12 V with a bias duration of 1 s. In addition to nanolines and nanopits, complex structures such as alphabets can be written on a polymer film.

The word "POLY" [Fig. 5(a)] was drawn with a line thickness of 78 nm by moving the tip at a speed of 0.5 μm/s using a voltage of -5 V. Similarly, a grooved pattern of "PMAA" with a linewidth of 195 nm was written at -12 V using a tip speed of 0.5 μm/s [Fig. 5(b)].

It is known that the formation of a raised pattern is due to AFM-tip-induced polymer melting and mass transport of softened polymers.[98–104] The indentation of a polymer film via AFM-tip-induced local Joule heating was demonstrated earlier using a thermomechanical writing process.[105] However, patterning was done at a voltage of -3 V, and the ablation of the polymer occurred at an increased voltage of -11 V. To explain the formation of raised and grooved patterns with an increase of applied bias on the PMAA film, a current–voltage (I–V) curve was recorded during the pattern formation by ramping the tip using various tip bias values at a scan rate of 1 Hz. The I–V curves [Fig. 6(a)] showed linear behavior at the low voltage region, but deviated from linearity at higher voltages of -9 or -11 V. The change in linearity at high-voltage ramping, i.e. the decrease in current, could be due to the depolymerization [Fig. 6(b)]. After depolymerization,

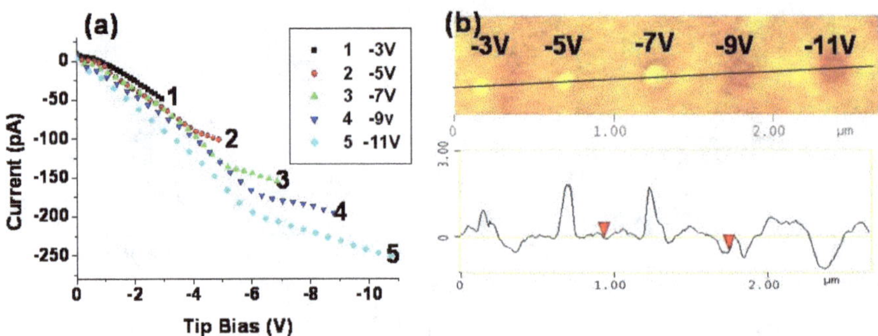

Fig. 6. (a) *I–V* curve of a PMAA film of thickness 60 nm by ramping the tip to various voltages at a scan rate of 1 Hz and (b) the corresponding raised and grooved patterns. (*Adapted from Ref. 99.*)

no raised features were observed around the groove pattern, which suggests that the ablation of the polymer film occurred at a high voltage. In addition, successive ramping of bias on the polymer film led to continuous removal of polymer film, which in turn increased the current due to a decrease in polymer thickness.

(b) Nanopatterning of poly(methacrylate) (PMA) polymer films

The effect of hydrophobicity/hydrophilicity of the polymer on water meniscus formation between the AFM tip and the substrate was investigated using a hydrophobic polymer (PMA) and a hydrophilic polymer (PAA). Figure 7(a) shows the line patterns formed at different applied biases, from -9 V to -12 V, with a tip speed of 0.1 μm/s, on PMA-coated Si substrate. During the experiments, it was observed that line patterning of the PMA polymer was not regular by forming continuous and discontinuous lines [Fig. 7(a)].

When lines were drawn on the PMA surface with a tip bias of -9 V, no line formation was observed. But the formation of line patterns increased slowly when the voltage was increased. Discontinuous lines were formed at -10 and -11 V, while a continuous line was formed at -12 V with the same tip speed (0.1 μm/s) of patterning. These results confirm that the applied voltage has a significant effect on the formation of continuous patterns. The groove patterning from PMA film may be attributed to the low T_g (6°C) value of the polymer.[99] But no significant change in the width of the pattern was observed with respect to the applied bias, which suggests that the width of the pattern is governed by the size of the water meniscus formed between

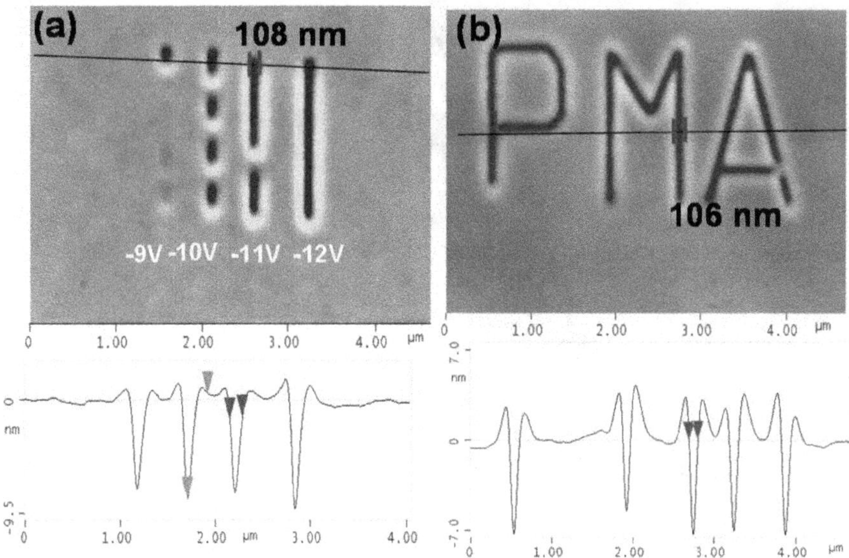

Fig. 7. (a) Nanolines drawn on the PMA film at a tip bias of −9, −10, −11 and −12 V, with a tip speed of 0.1 μm/s. (b) Continuous, grooved structure drawn at −12 V, at a tip of speed 0.1 μm/s.

the tip and the substrate. Also, patterns were not observed up to a tip voltage of −8 V. This clearly indicates that PMA can be patterned at −9 V whereas in the PMAA film the pattern formation starts at a higher voltage, −18 V.

In order to ascertain whether the discontinuous line formation was due to the effect of tip voltage or tip speed, patterning on PMA film was done at different tip speeds, with a constant applied bias. Pentagonal structure was drawn on the PMA film at two different tip speeds of 0.25 and 0.1 μm/s, with a tip bias of −12 V, and the results are shown in Fig. 8(a). The pentagon pattern width of 108 nm was obtained at a tip bias of −12 V. Here, the regular pattern was observed at a tip speed of 0.1 μm/s rather than an increased tip speed of 0.25 μm/s. Since PMA readily forms nanodots rather than nanolines, an array of nanodots was patterned at a tip bias of −9 V with a tip contact time of 2 s [Fig. 8(b)].

(c) Nanopatterning of polyacrylic acid (PAA) polymer film

In addition, the influence of hydrophobicity or hydrophilicity of the polymers on the nanopatterning was examined using hydrophilic PAA. Figure 9(a) shows that nanolines can be drawn on PAA-coated Si(100) substrate with different voltages varying from −5 V to −11 V at an AFM

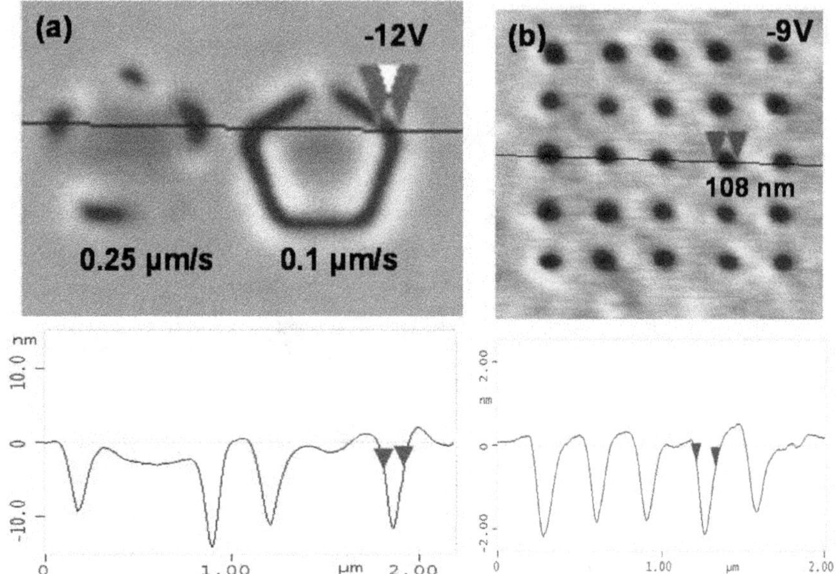

Fig. 8. (a) Pentagon pattern drawn on the PMA film at a tip speed of 0.25 μm/s, 0.1 μm/s with a tip bias of −12 V. (b) Array of nanogroove pattern with a size of 108 nm fabricated at −9 V with a tip contact time of 2 s. (*Adapted from Ref. 106.*)

tip speed of 0.5 μm/s, and the observed line widths were 39, 59, 117 and 137 nm, respectively. Here, grooved patterns were formed at all voltages and the formation of pattern width and depth was increased with an increase in applied voltage.

In order to find the effect of tip speed on patterning, PAA films were patterned with different tip speeds of 1, 0.5, 0.25 and 0.1 μm/s at a constant tip bias of −7 V, and the corresponding linewidths were 43, 56, 71 and 84 nm, respectively [Fig. 9(b)]. These results show that both the applied voltage and the tip speed have a significant effect on the pattern formation, as expected. In addition to the line pattern, the abbreviation "PAA" was drawn on the polymer film with a pattern width of 59 nm at a tip bias of −7 V and a tip speed of 0.5 μm/s [Fig. 10(a)]. Also, an array of dot patterns was drawn at a tip voltage of −8 V with a tip contact time of 1 s [Fig. 10(b)]. Compared to the PMA film, the array of dot patterns was formed at a smaller tip contact time of 1 s, and at a lower tip voltage of −8 V.

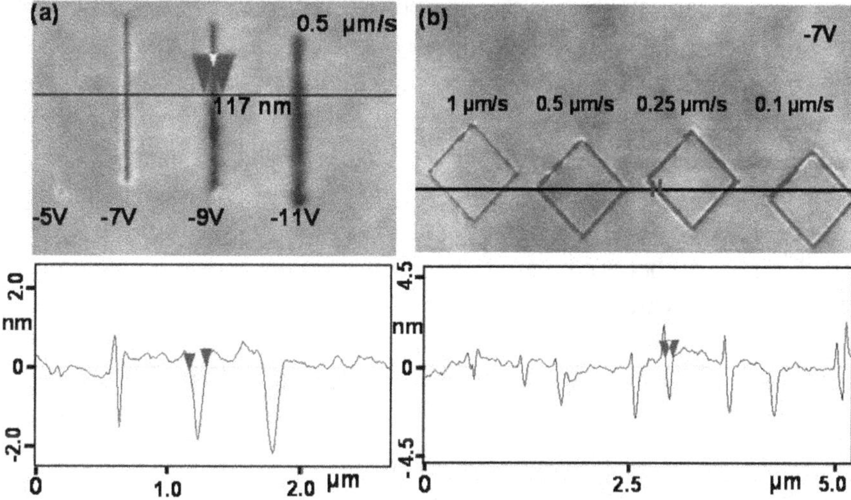

Fig. 9. (a) Nanolines written on the PAA film at a constant tip speed of 0.5 μm/s with a different bias of -5, -7, -9 and -11 V, and the corresponding linewidths 35, 59, 117 and 137 nm, respectively. (b) AFM image of square patterning of the PPA film drawn at -7 V with a tip speed of 1, 0.5, 0.25 and 0.1 μm/s. (*Adapted from Ref. 106.*)

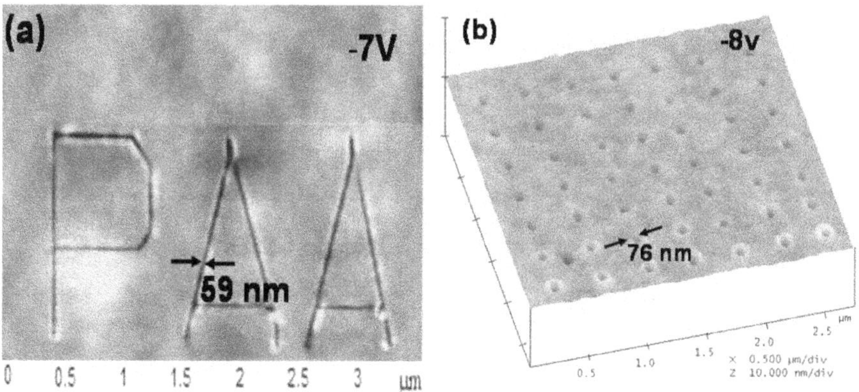

Fig. 10. (a) Nanolines of the abbreviation "PAA" with a 59 nm width formed at a tip bias of -7 V and a tip speed of 0.5 μm/s. (b) Array of grooved patterns of 78 nm width drawn on the PAA film at a tip voltage of -8 V and a tip contact time of 1 s. (*Adapted from Ref. 106.*)

A significant structural difference between PMA and PAA involves the presence of –COOH groups along the polymer backbone of PAA. Moreover, the –COOH groups on the PAA backbone are more electron-withdrawing than the $-CO_2Me$ groups in PMA. So, it is conceivable that interchain hydrogen bonding as well as depolymerization of the polymer backbone in PAA may facilitate the formation of nanopatterns. Under ambient conditions, water condenses at the tip and forms a bridge with the polymer film.[108] It is also believed that the polar acid groups in the polymer chain interact strongly with water and enhance the ionic conductivity of the film. The applied voltage between the tip and the substrates creates effective Joule heating[98] to raise the temperature of the polymer film that causes the ablation of the polymer backbone, which results in the formation of groove-type patterns (Fig. 10).

It is conceivable that the PAA requires only low voltage as compared to PMA for nanopatterning, due to the presence of a large number of acidic groups on the polymer backbone, which enhances the ionic conductivity as well as depolymerization of the film.

10.1.4. *Electrochemical nanolithography*

Electrochemical nanolithography is a lithographic technique for patterning the polymer/organic substrate through electrochemical methods by using a biased AFM tip. In this method, a water meniscus formed between AFM tip and substrate was used as an electrochemical cell for electrochemical changes. A schematic representation of the nanopatterning of polymer film by the electrochemical oxidation method is shown in Fig. 11. In this process, a negative bias voltage was applied to the AFM tip (cathode), and the electrochemical oxidation occurs on the polymer/organic-coated conducting electrode (acting as anode). This technique is used to fabricate nanostructures by the oxidation and cross-linking of the precursor electroactive polymer, such as polyvinylcarbazole (PVK).

10.1.4.1. *Nanopatterning of PVK films*

Nanopatterning of PVK polymer films deposited on Si substrate was performed at various tip speed and biases. Figure 12(a) shows that protruding nanopatterned lines were formed on the PVK film when the negatively

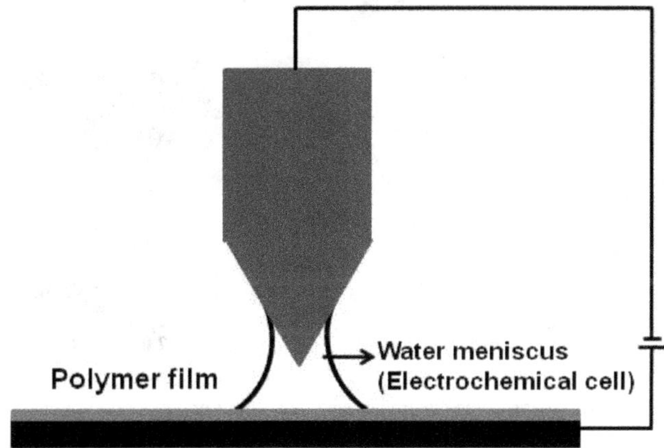

Fig. 11. Schematic representation of the nanopatterning of polymer film by the electrochemical nanolithography method.

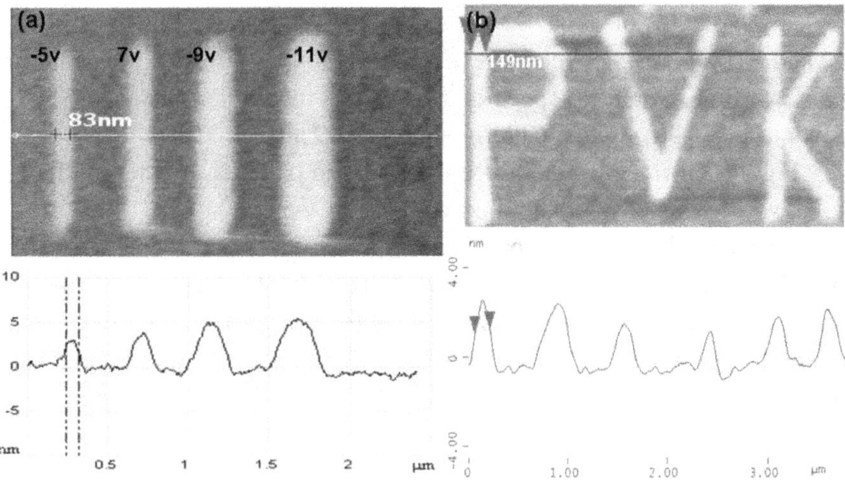

Fig. 12. (a) Nanolines written on the PVK film at a constant tip speed of 1 μm/s with a different bias of −5, −7 and −9 V and corresponding to linewidth of 83, 128, 162 and 231 nm, respectively. (b) AFM image with a height profile of the pattern "PVK" drawn at −7 V at a tip speed of 1 μm/s. (*Adapted from Ref. 101.*)

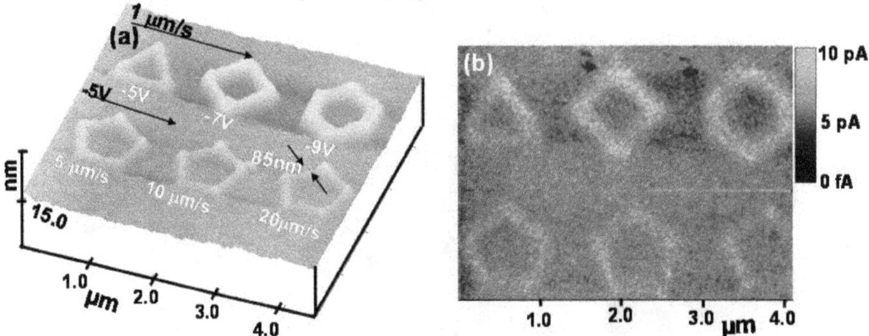

Fig. 13. (a) AFM height image of the electrochemical patterning of a polygon at different tip voltages and speeds. (b) Corresponding C-AFM image of polygon patterning with conductive current 10 pA. (*Adapted from Ref. 101.*)

biased AFM tip was moved on top of the polymer film. The nanolines were formed at voltages of −5, −7, −9 and −11 V with linewidths of 83, 128, 162 and 231 nm, respectively. The height (topological) image of the pattern "PVK" written with a tip speed of 1 μm/s and an applied bias of −7 V is shown in Fig. 12(b) with a pattern width of 149 nm.

Figure 13(a) shows that different polygon patterns grown on a PVK film at various speeds with applied voltages of −5 and −7 V.

These results show that patterns of specific sizes and shapes could be drawn by controlling the applied voltage and tip bias. Due to the electropolymerization of PVK via cross-linking between the carbazole moieties, the patterned area should be conductive. In order to check the conductivity of this cross-linked pattern on the Si(100) substrate, the current mapping was done by scanning the patterned surface with a conductive AFM tip at a sample bias of −5 V. The current mapping image of the polygon pattern is shown in Fig. 13(b), in which the pattern exhibiting a conductivity of 10 pA was observed. In another experiment of I–V curve measurements, we found that the current reaches a saturation conductivity of 142 nA at the bias of 8 V when ramping the voltage from 0 to 12 V through the AFM tip. No conductivity was observed at a nonpatterned area, which indicates that electro-cross-linking between the carbazole units on the polymer backbone leads to the conductivity of the pattern.

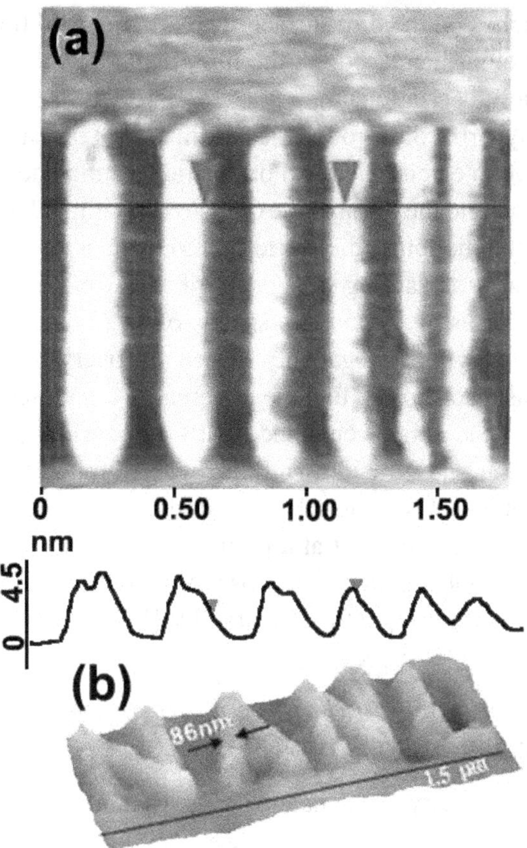

Fig. 14. (a) Nanopatterns drawn on carbazole film at a constant bias of −7 V at various tip speeds of 1, 2, 4, 6, 8 and 10 μm/s, corresponding to linewidths of 187, 162, 140, 128, 87 and 78 nm, respectively, and imaged by contact mode AFM (height). (b) 3D image of carbazole monomer patterned at −7 V with a pattern width of 86 nm. The height is at ∼2 nm. (*Adapted from Ref. 101.*)

10.1.4.2. *Nanopatterning of carbazole monomer*

Using the spin-coated carbazole monomer film, polymerized nanopatterns were generated on a Si substrate with a negatively biased tip. Figure 14(a) shows the contact mode AFM image of the nanopatterned monomer film on a Si substrate. A clear image could be obtained when the patterned surface was scanned with an AFM tip at a relatively weak contact force of 2–5 nN. When the contact force was raised, the AFM image showed an increased number of scratches on the film surface. The polymerized lines were characterized by a height increase over the film, owing to the applied current and the electropolymerization process. The electropolymerization led to the formation of polycarbazole through the 3,6-linkage of the carbazole groups. The formation of an electrochemical bridge between the negatively biased AFM tip (cathode) and the Si substrate (anode) is facilitated by the influence of humidity on the monomer dielectric film to form counterions.[102]

Polymer lines of varying linewidths of 187, 162, 140, 128, 87 and 78 nm were drawn using various tip speeds of 1, 2, 4, 6, 8 and 10 μm/s, respectively, by moving a negatively biased tip over the monomer film [Fig. 14(a)]. This also showed that the pattern size decreased with increasing tip speed, but the height remained relatively constant ($\sim$2 nm) during the patterning. The pattern width and height were found to increase with increasing tip voltage. The pattern height varied only when the voltage was changed and remained constant at a particular tip bias. In addition to the nanoline, other complex patterns such as the word "NANO," with a width of 86 nm, were drawn using a tip bias of −7 V [Fig. 14(b)].

10.1.4.3. *Conductive and thermal properties of patterned films*

To study the conductivity of the nanopattern, a square area of the film was patterned by scanning the negatively biased AFM tip across a 1 μm^2 area with a scan speed of 1 Hz at −5 V. Figure 15(a) shows the protruding square pattern of an electropolymerized carbazole (forming conjugated polycarbazole) with a height of 5 nm. The corresponding C-AFM image (scanned at a positive bias, +5 V) [Fig. 15(b)] shows that the patterned area is more conductive with a current of 10 pA compared to the unpatterned area. This C-AFM experiment is somewhat similar to STM measurements where the topographic image is formed because of a constant tunneling current. In the case of C-AFM, a current may tunnel from the AFM tip to the nonconducting polymer film during scanning (tip voltage: +5 V) such that the current mapped image shows that the current is much lower for the unpatterned area than for the patterned area. This change can be clearly seen from the image where the patterned area is more conductive than the unpatterned area by C-AFM. The thermal stability of the patterns was also investigated by heating the patterned substrate for 3 h at two different temperatures, 150 and

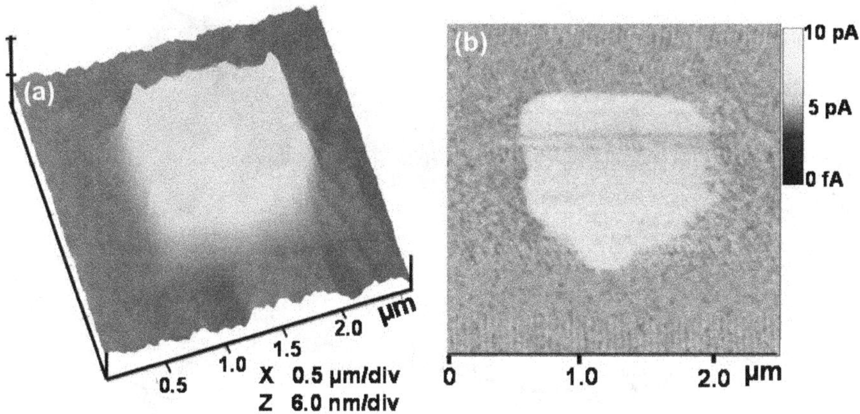

Fig. 15. (a) Square pattern of size 1 μm^2 on carbazole film at a scanning speed of 1 Hz with a tip bias of −5 V imaged by contact mode AFM. (b) The corresponding current mapping image (C-AFM) of the conductive square pattern with a conducting current of 10 pA at an applied bias of +5 V for imaging. (*Adapted from Ref. 101.*)

270°C. At both temperatures, the patterns were stable and the conductivity did not change after heating.

The pattern shown in Fig. 16 represents a feature annealed at 270°C for 3 h and then cooled to room temperature. Even though the temperature is above the melting point (245°C) of carbazole, the thickness or aspect ratio of the patterns did not change after annealing. In this case, the formation of a high T_g and T_m linear polycarbazole prevented thermal collapse of the nanopatterns.

10.1.4.4. *Nanopatterning of electroactive copolymer film*

Figure 18(a) shows the formation of coronas on a spin-coated polymer film (polymer A) through electrochemical process (Fig. 17) with diameters of 658, 2778 and 5320 nm based on tip-to-substrate biased voltages of −7, −9 and −11 V, respectively. Detailed information on the synthesis of the polymers poly([3-{2-(4-vinylphenoxy)ethyl}thiophene]-*co*-9-[2-(4-vinylphenoxy)ethyl]-9*H* carbazole) (PMTC) and conducting poly([3-{2-(4-vinylphenoxy)ethyl}thiophene]-*co*-9-[2-(4-vinylphenoxy) ethyl]-9*H*-carbazole) (CPMTC) is given elsewhere.[107] The patterning was done by holding the AFM tip for 2 s on the polymer film. Similarly, Fig. 18(b) shows the lines drawn on the polymer film at different voltages

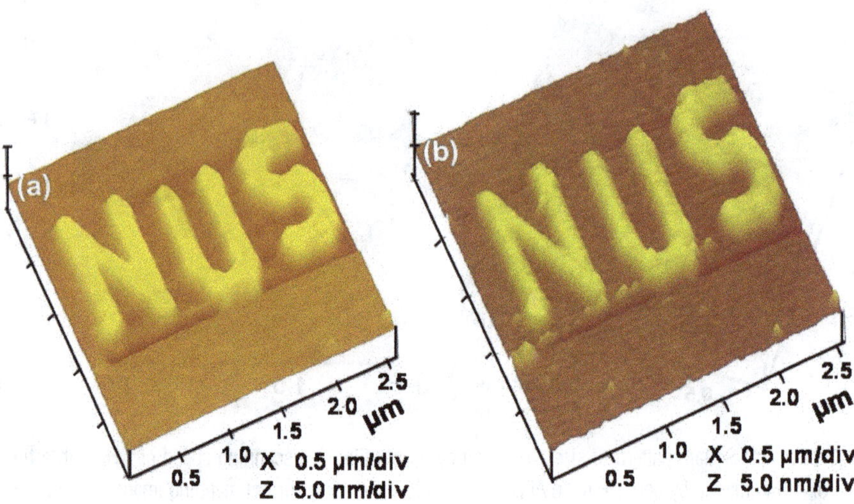

Fig. 16. AFM height image of the patterned abbreviation "NUS" before (a) and after (b) heating at 270 °C for 3 h of the carbazole film. (*Adapted from Ref. 101.*)

Fig. 17. Electropolymerization of neutral precursor polymer A (PMTC) to cross-linked conducting polymer B (CPMTC).[107]

of −7, −9 and −11 V using a tip speed of 1 μm/s. In these experiments, the dimensions of the patterned areas were observed to increase enormously with increasing applied bias (negative). The size of the dot pattern is larger at higher voltages (−9 and −11 V). The same effect can be seen on the lines drawn at higher voltages such as −11 V. This observation indicates that with increased voltage the patterning rate of the polymer film was

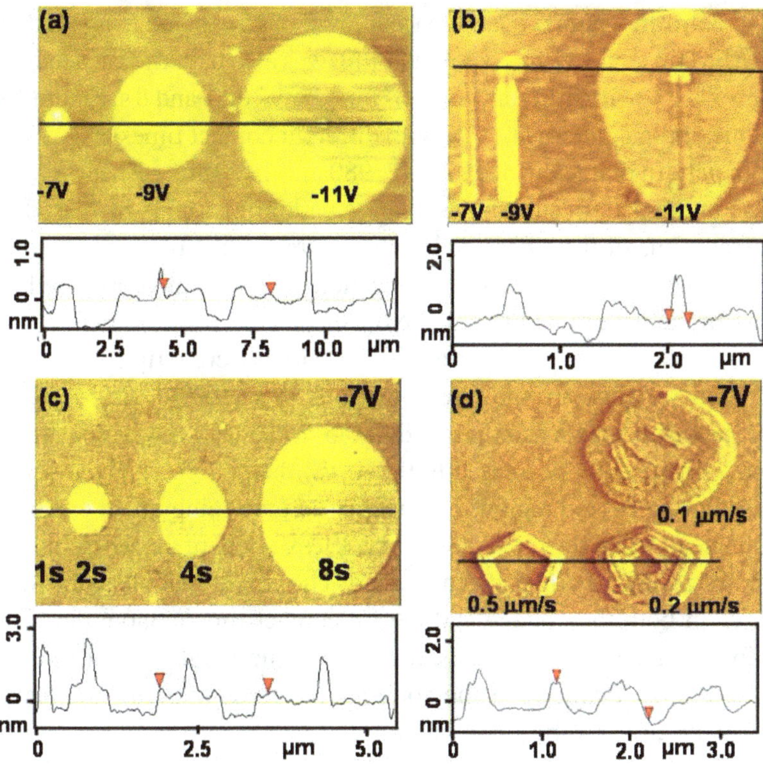

Fig. 18. (a) Dot pattern of polymer (A) film at various biases of −7, −9 and −11 V with a tip contact time of 2 s and corresponding diameters of 658, 2778 and 5320 nm, respectively. (b) Line pattern of polymer A at a constant tip speed of 1 μm/s with different applied voltages of −7, −9 and −11 V. (c) Dot patterns drawn on polymer film with a constant bias of −7 V at various tip contact times of 1, 2, 4 and 8 s and corresponding pattern widths of 190, 642, 980 and 1990 nm, respectively. (d) Height image of the hexagon pattern of the polymer at −7 V with tip speeds of 0.5, 0.2 and 0.1 μm/s. (*Adapted from Ref. 108.*)

increased, which led to corona-type patterns with increased pattern width. In this case, the only difference in the patterning condition between the dot pattern and the line pattern was the tip speed i.e. the time for which the tip stays for a certain location on the polymer.

In the dot pattern the tip was positioned on the film for 2 s, whereas in the line pattern it was moved over the film at a speed of 1 μm/s. This clearly indicates the importance of the electric field distribution in electrochemical nanopatterning procedures, which has not been observed in DPN-type nanopatterning.

To confirm whether the increase in pattern size is due to applied voltage or AFM tip contact time, a polymer film was also patterned at a fixed low voltage of −7 V with different contact times between 1 and 8 s. Figure 18(c) shows the corona pattern at −7 V with different contact times of 1, 2, 4 and 8 s, with pattern diameters of 190, 642, 980 and 1990 nm, respectively. The increase in the pattern size shows that the patterning rate is increased with the contact time of the AFM tip. In other words, when the applied voltage time is increased, an increase in the pattern width is also observed. The increase in the pattern width with the tip speed is also shown in Fig. 18(d), where hexagons have been drawn at −7 V with different tip speeds.

When we compare the three hexagons in Fig. 18(d), it can be seen that the hexagon drawn at a tip speed of 0.5 μm/s has a regular pattern, but when the tip speed is decreased to 0.2 μm/s, the pattern changes with the increased polymerized region. This increased patterning effect can be further enhanced when the tip is moved at 0.1 μm/s, where the whole polymer film close to the AFM tip end is patterned. These observations indicate that increases in both the voltage and the tip contact time enlarge the pattern size. To understand the relationship between the height (relief structure, Fig. 17) and the diameter of the corona from these patterns, the features were formed at −7 V by immobilizing the AFM tip for 5 s, as shown in

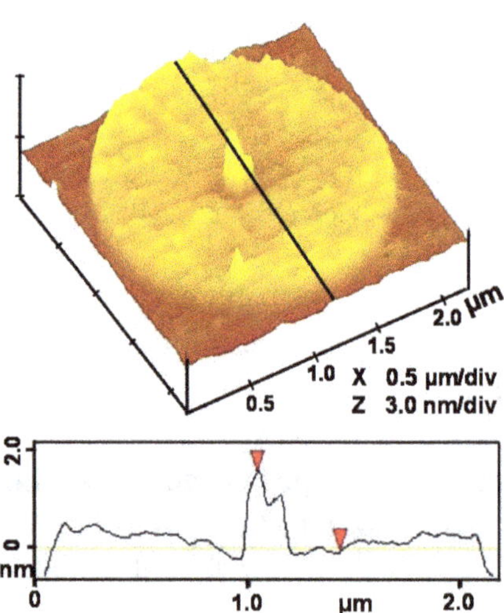

Fig. 19. Three-dimensional nanostructure of a polymer corona pattern and the corresponding height profile formed at −7 V with a tip contact of 6 s. (*Adapted from Ref. 108.*)

Fig. 18. It appears that the pattern height at the center is higher than in other areas, indicating that the polymer under the tip is more exposed to the applied voltage, which causes protruded areas at the center rather than in other areas. Recently, the distribution of the electric field in a C-AFM experiment for patterning polymer films has been modeled.[99] Also, the formation of corona patterns on a dielectric film at high voltage has been reported using a pin-plane corona model.[109,110]

The high charge concentration under the AFM tip due to the strong electric field leads to the central protruding pattern formation, and the localized discharge of electric fields forms the corona pattern through the electropolymerization process. It has been reported that the formation of a corona-type oxide pattern on a Si substrate at high voltage (−45 V) occurs because of electronic conduction rather than ionic conduction.[109] However, in this case a Si substrate was coated with an electroactive polymer and the formation of a corona-type pattern at a lower voltage (<12 V) was observed. The rapid patterning of this polymer is quite different from that of other previously investigated homopolymers such as PVK and polythiophene, where corona-type patterns were not observed.[101,108] The maximum linewidth obtained by patterning the polymer PVK under an applied bias of −11 V with a tip speed of 1 μm/s was 182 nm.[101]

10.2. Applications and Challenges of AFM Nanolithography

The use of AFM in nanoscale fabrication of surfaces has been accelerated with the development of nanografting techniques in 1997.[25] In nanografting, an AFM tip is used to create a pattern in an organic SAM formed on a gold surface. A force is applied to a selected area of the film to remove the adsorbed thiol molecules. A different thiol is adsorbed to the exposed gold to form a nanoscale feature on the surface. Gold–thiol SAMs (thin, uniform and chemically stable monolayers) provided a promising platform for constructing chemically specific nanostructures. By introducing proper terminal functional groups, the surface properties of the SAMs can be modified.

In another application, a scanning-probe-based data storage concept was developed by IBM researchers using a heated AFM probe. In this thermomechanical writing process, an AFM probe writes a data bit over a polymer surface through combined heat and mechanical force of the tip, which causes the polymer to soften, thereby facilitating the writing. Information is stored as sequences of "indentation" and "no indentation" written on polymer films using an array of AFM cantilevers.[45–49] This probe-based data storage combines ultrahigh density, small form factor and high data rates by means of highly parallel operation of a large number of probes.

The primary advantage of AFM lithography in nanofabrication by the use of scanning probes has been discussed in several reviews.[12–19] However, there are two important technological issues that should be resolved in order to get rid of the limitations of this method. First, the writing speed of the AFM lithography systems is too slow for mass production because of the serial patterning process. The typical writing speed for a single tip is about 2 μm/s. Researchers are making a huge effort to overcome this drawback. The aim is to produce an array of multiple tips and to write using all the tips simultaneously, thereby increasing the speed. It has been reported that anisotropic wet etching can make arrays of thousands of cantilevers. The second obstacle involves the friction-induced wear of the SPM tip during fabrication. One potential solution to this problem may be the use of nanotube probes, which are very robust and do not increase in diameter during wear owing to the cylindrical geometry. Such probes will enable us to write extremely thin patterns of several-nanometer width without noticeable wear. In spite of speed and wear problems, the use of SPM families in nanoscience has been increasing continuously. Some of the challenges, such as understanding the mechanism of structure formation from various materials, chemical characterization of nanoscale structure and ultimate resolution of the pattern formation, need to be explored through careful studies. From the perspectives of low cost, operational simplicity and advancing scientific investigation, SPL has much promise and opens the door to exploration of many areas where high-resolution photolithography and particle beam writing are not applicable.

References

1. W. M. Moreau, in *Semiconductor Lithography: Principles and Materials* (Plenum, New York, 1988).

2. D. Brambley, B. Martin and P. D. Prewett, *Adv. Mater. Opt. Electron.* **4**, 55 (1994).
3. A. M. Goethals, P. D. Bisschop, J. Hermans, R. Jonckheere, F. Van Roey, D. Van den Heuvel, A. Eliat and K. Ronse, *J. Photopolym. Sci. Technol.* **16**, 549 (2003).
4. J. Mulkens, J. McClay, B. Tirri, M. Brunotte, B. Mecking and H. Jasper, *Proc. SPIE–Int. Soc. Opt. Micro.* **5040**, 753 (2003).
5. F. Cerrina, S. Bollepalli, M. Khan, H. Solak, W. Li and D. He, *Microelectron. Eng.* **53**, 13 (2000).
6. R. L. Brainard, J. Cobb and C. A. Cutler, *J. Photopolym. Sci. Technol.* **16**, 401 (2003).
7. F. Schellenberg, *IEEE Spectrum* **9**, 34 (2003).
8. M. J. Lercel, G. F. Redinbo, F. D. Pardo, M. Rooks, R. C. Tiberio, P. Simpson, H. G. Craighead, C. W. Sheen, A. N. Parikh and D. L. Allara, *J. Vac. Sci. Technol.* **B12**, 3663 (1994).
9. M. J. Lercel, H. G. Craighead, A. N. Parikh, K. Seshadri and D. L. Allara, *Appl. Phys. Lett.* **68**, 1504 (1996).
10. A. Pique and D. B. Chrisey, in *Direct Write Technology for Rapid Prototype Application* (Academic, New York, 2002).
11. C. F. Quate, *Surf. Sci.* **386**, 259 (1997).
12. N. Mino, S. Ozaki, K. Ogawa and M. Hatada, *Thin Solid Films* **243**, 374 (1994).
13. R. M. Nyffenegger and R. M. Penner, *Chem. Rev.* **97**, 1195 (1997).
14. S. Kondo, S. Heike, M. Lutwyche and Y. Wada, *J. Appl. Phys.* **78**, 155 (1995).
15. A. Majumdar and S. M. Lindsay, in *Technology of Proximal Probe Lithography*; (SPIE — The International Society for Optical Engineering, Bellingham, Washington, 1993), p. 33.
16. C. R. K. Marrian, E. A. Dobisz and J. A. Dagata, in *Technology of Proximal Probe Lithography* (SPIE — The International Society for Optical Engineering, Bellingham, Washington, 1993), p. 58.
17. D. Wouters and U. S. Schubert, *Angew. Chem. Int. Ed.* **43**, 2480 (2004).
18. D. S. Ginger, H. Zhang and C. A. Mirkin, *Angew. Chem. Int. Ed.* **43**, 30 (2004).
19. B. D. Gates, Q. Xu, M. Stewart, D. Ryan, C. G. Willson and G. M. Whitesides, *Chem. Rev.* **105**, 1171 (2005).
20. K. Wadu-Mesthrige, S. Xu, N. A. Amro and G. Liu, *Langmuir* **15**, 8580 (1999).
21. N. A. Amro, S. Xu and G. Liu, *Langmuir* **16**, 3006 (2000).
22. K. J. Schoer and R. M. Crooks. *Langmuir* **13**, 2323 (1997).
23. J. Wang, J. R. Kenseth, V. W. Jones, J. B. Green, M. T. McDermott and M. D. Porter, *J. Am. Chem. Soc.* **119**, 12796 (1997).
24. B. Cappella and H. Sturm, *J. Appl. Phys.* **91**, 50 (2002).
25. S. Xu and G. Y. Liu, *Langmuir* **13**, 127 (1997).
26. S. Xu, S. Miller, P. E. Laibinis and G. Y. Liu, *Langmuir* **15**, 7244 (1999).
27. Y. Sugimoto, M. Abe, S. Hirayama, N. Oyabu, O. Custance and S. Morita, *Nat. Mater.* **4**, 156 (2005).
28. K. Salaita, Y. Wang and C. A. Mirkin, *Nat. Nanotechnol.* **2**, 145 (2007).

29. E. S. Snow and P. M. Campbell, *Appl. Phys. Lett.*, **64**, 1932 (1994).
30. Y. Li, B. W. Maynor and J. Liu, *J. Am. Chem. Soc.* **123**, 2105 (2001).
31. S. F. Lyuksyutov, R. A. Vaia, P. B. Paramonov, S. Juhl, L. Waterhouse, R. M. Ralich, G. Sigalov and E. Sancaktar, *Nat. Mater.* **2**, 468 (2003).
32. W. P. King, T. W. Kenny, K. E. Goodson, G. Cross, M. Despont, U. Dürig, H. Rothuizen, G. K. Binnig and P. Vettiger, *Appl. Phys. Lett.* **78**, 1300 (2001).
33. B. Cappella, H. Sturm and S. M. Weidner, *Polymer* **43**, 4461 (2002).
34. T. A. Jung, A. Moser, H. J. Hug, D. Brodbeck, R. Hofer, H. R. Hidber and U. D. Schwarz, *Ultramicroscopy* **42**, 1446 (1992).
35. X. Jin and W. N. Unertl, *Appl. Phys. Lett.* **61**, 657 (1992).
36. M. Heyde, K. Rademann, B. Cappella, M. Geuss, H. Sturm, T. Spangenberg and H. Niehus, *Rev. Sci. Instrum.* **72**, 136 (2001).
37. H.-Y. Nie, M. Motomatsu, W. Mizutani and H. Tokumoto, *J. Vac. Sci. Technol. B* **13**, 1163 (1995).
38. B. D. Terris, S. A. Rishton, H. J. Mamin, R. P. Reid and D. Rugar, *Appl. Phys. A* **66**, S809 (1998).
39. H. J. Mamin and D. Rugar, *Appl. Phys. Lett.* **61**, 1003 (1992).
40. R. P. Ried, H. J. Mamin, B. D. Terris, L. S. Fan and D. Rugar, *J. Microelectromech. Syst.* **6**, 294 (1997).
41. H. J. Mamin, R. P. Reid, B. D. Terris and D. Rugar, *Proc. IEEE* **87**, 1014 (1999).
42. P. Vettiger, G. Cross, M. Despont, U. Drechsler, U. Durig, B. Gotsmann, W. Haberle, M. A. Lantz, H. E. Rothuizen, R. Stutz and G. K. Binnig, *IEEE Trans. Nanotechnol.* **1**, 39 (2002).
43. M. I. Lutwyche, M. Despont, U. Drechsler, U. Durig, W. Haberle, H. Rothuizen, R. Stutz, R. Widmer, G. K. Binnig and P. Vettiger, *Appl. Phys. Lett.* **77**, 3299 (2000).
44. M. Lutwyche, C. Andreoli, G. Binnig, J. Brugger, U. Drechsler, W. Haberle, H. Rohrer, H. Rothuizen, P. Vettiger, G. Yaralioglu and C. Quate, *Sensors Actuators A Phys.* **73**, 89 (1999).
45. M. Despont, J. Brugger, U. Drechsler, U. Durig, W. Haberle, M. Lutwyche, H. Rothuizen, R. Stutz, R. Widmer, H. Rohrer, G. Binnig and P. Vettiger, *IEEE Int. MEMS Tech. Digest* 564 (1999).
46. U. Durig, G. Cross, M. Despont, U. Drechsler, W. Haberle, M. I. Lutwyche, H. Rothuizen, R. Stutz, R. Widmer, P. Vettiger, G. K. Binnig, W. P. King and K. E. Goodson, *Tribol. Lett.* **9**, 25 (2000).
47. P. Vettiger, J. Brugger, M. Despont, U. Drechsler, U. Durig, W. Haberle, M. Lutwyche, H. Rothuizen, R. Stutz, R. Widmer and G. Binnig, *J. Microelectron Eng.* **46**, 11 (1999).
48. M. Despont, J. Brugger, U. Drechsler, U. Dürig, W. Häberle, M. Lutwyche, H. Rothuizen, R. Stutz, R. Widmer, G. Binnig, H. Rohrer and P. Vettiger, *Sensors Actuators A Phys.* **80**, 100 (2000).
49. R. D. Piner, J. Zhu, F. Xu, S. Hong and C. A. Mirkin, *Science* **283**, 661 (1999).

50. S. Xu, S. Miller, P. E. Laibinis and G. Liu, *Langmuir* **15**, 7244 (1999).
51. G.-Y. Liu and M. B. Salmeron, *Langmuir* **10**, 367 (1994).
52. X. D. Xiao, G. Y. Liu, D. H. Charych and M. Salmeron, *Langmuir* **11**, 1600 (1995).
53. S. O. Kelley, J. K. Barton, N. M. Jackson, L. D. McPherson, A. B. Potter, E. M. Spain, M. J. Allen and M. G. Hill, *Langmuir* **14**, 6781 (1998).
54. D. Zhou, K. Sinniah, C. Abell and T. Rayment, *Langmuir* **18**, 8278 (2002).
55. J. W. Zhao and K. Uosaki, *Langmuir* **17**, 7784 (2001).
56. H. Sugimura, K. Okiguchi, N. Nakagiri and M. Miyashita, *J. Vac. Sci. Technol. B.* **14**, 4140 (1996).
57. H. Sugimura and N. Nakagiri, *J. Vac. Sci. Technol. B.* 15, 1394 (1997).
58. H. Sugimura and N. Nakagiri, *Nanotechnology* **8**, A15 (1997).
59. H. Sugimura and N. Nakagiri, *J. Am. Chem. Soc.* **119**, 9226 (1997).
60. D. S. Cruchon, S. Porthun and G. Y. Liu, *Appl. Surf. Sci.* **175**, 636 (2001).
61. S. Xu, N. A. Amro and G. Y. Liu, *Appl. Surf. Sci.* **175**, 649 (2001).
62. M. Z. Liu, N. A. Amro, C. S. Chow and G. Y. Liu, *Nano Lett.* **2**, 863 (2002).
63. M. A. Case, G. L. McLendon, Y. Hu, T. K. Vanderlick and G. Scoles, *Nano Lett.* **3**, 425 (2003).
64. N. A. Amro, S. Xu and G. Y. Liu, *Langmuir* **16**, 3006 (2000).
65. J. C. Garno, Y. Yang, N. A. Amro, S. D. Cruchon, S. Chen and G. Y. Liu, *Nano Lett.* **3**, 389 (2003).
66. P. V. Schwartz, *Langmuir* **17**, 5971 (2001).
67. K. M. Wadu, S. Xu, N. A. Amro and G. Y. Liu, *Langmuir* **15**, 8580 (1999).
68. J. R. Kenseth, J. A. Harnisch, V. W. Jones and M. D. Porter, *Langmuir* **17**, 4105 (2001).
69. J. F. Liu, D. S. Cruchon, J. C. Garno, J. Frommer and G. Y. Liu, *Nano Lett.* **2**, 937 (2002).
70. H. Sugimura and N. Nakagiri, *J. Vac. Sci. Technol. A* **14**, 1223 (1996).
71. H. J. Mamin and D. Rugar, *Appl. Phys. Lett.* **61**, 1003 (1992).
72. R. P. Ried, H. J. Mamin, B. D. Terris, L. S. Fan and D. Rugar, *J. Microelectromech. Syst.* **6**, 294 (1997).
73. H. J. Mamin, R. P. Reid, B. D. Terris and D. Rugar, *Proc. IEEE* **87**, 1014 (1999).
74. P. Vettiger, G. Cross, M. Despont, U. Drechsler, U. Durig, B. Gotsmann, W. Haberle, M. A. Lantz, H. E. Rothuizen, R. Stutz and G. K. Binnig, *IEEE Trans. Nanotechnol.* **1**, 39 (2002).
75. P. J. Thomas, G. U. Kulkarni and C. N. R. Rao, *J. Mater. Chem.* **14**, 625 (2004).
76. J. H. Lim, D. S. Ginger, K. B. Lee, J. Heo, J. M. Nam and C. A. Mirkin, *Angew. Chem. Int. Ed.* **42**, 2309 (2003).
77. J. M. Nam, S. W. Han, K. B. Lee, X. Liu, M. A. Ratner and C. A. Mirkin, *Angew. Chem. Int. Ed.* **43**, 1246 (2004).
78. L. M. Demers, D. S. Ginger, S.-J. Park, Z. Li, S.-W. Chung and C. A. Mirkin, *Science* **296**, 1836 (2002).

79. L. Ding, Y. Li, H. Chu, X. Li and J. Liu, *J. Phys. Chem. B.* **109**, 22337 (2005).
80. S. Hong, J. Zhu and C. A. Mirkin, *Science* **286**, 523 (1999).
81. A. Ivanisevic and C.A. Mirkin, *J. Am. Chem. Soc.* **123**, 7887 (2001).
82. A. Ivanisevic, J. H. Im, K. B. Lee, S. J. Park, L. M. Demers, K. J. Watson and C. A. Mirkin, *J. Am. Chem. Soc.* **123**, 12424 (2001).
83. X. Liu, S. Guo and C. A. Mirkin, *Angew. Chem. Int. Ed.* **42**, 4785 (2003).
84. D. L. Wilson, R. Martin, S. Hong, M. G. Cronin, C. A. Mirkin and D. L. Kaplan, *Proc. Natl. Acad. Sci. U.S.A.* **98**, 13660 (2001).
85. B. W. Maynor, Y. Li and J. Liu, *Langmuir* **17**, 2575 (2001).
86. G. Gundiah, N. S. John, P. J. Thomas, G. U. Kulkarni, C. N. R. Rao and S. Heun, *Appl. Phys. Lett.* **84**, 5341 (2004).
87. L. Fu, X. Liu, Y. Zhang, V. P. Dravid and C. A. Mirkin, *Nano Lett.* **3**, 757 (2003).
88. M. Su, X. Liu, S. Y. Li, V. P. Dravid and C. A. Mirkin, *J. Am. Chem. Soc.* **124**, 1560 (2002).
89. J. R. Hampton, A. A. Dameron and P. S. Weiss, *J. Phys. Chem. B.* **109**, 23118 (2005).
90. J. R. Hampton, A. A. Dameron and P. S. Weiss, *J. Am. Chem. Soc.* **128**, 1648 (2006).
91. J. Jang, G. C. Schatz and M. A. Ratner, *J. Chem. Phys.* **116**, 3875 (2002).
92. J. Jang, S. Hong, G. C. Schatz and M. A. Ratner, *J. Chem. Phys.* **115**, 2721 (2001).
93. N. Cho, S. Ryu, B. Kim, G. C. Schatz and S. Hong, *J. Chem. Phys.* **124**, 024714 (2006).
94. M. B. Ali, T. Ondarcuhu, M. Brust and C. Joachim, *Langmuir* **18**, 872 (2002).
95. S. F. Lyuksyutov, R. A. Vaia, P. B. Paramonov, S. Juhl, L. Waterhouse, R. M. Ralich, G. Sigalov and E. Sancaktar, *Nat. Mater.* **2**, 468 (2003).
96. S. F. Lyuksyutov, P. B. Paramonov, S. Juhl and R. A. Vaia, *Appl. Phys. Lett.* **83**, 4405 (2003).
97. E. Schaffer, T. A. Thurn, T. P. Russell and U. Steiner, *Nature* **403**, 874 (2000).
98. S. Jegadesan, C. A. Rigoberto and S. Valiyaveettil, *Adv. Mater.* **17**, 1282 (2005).
99. S. Jegadesan, S. Sindhu and S. Valiyaveettil, *Small* **2**, 481 (2006).
100. P. Vettiger, G. Cross, M. Despont, U. Drechsler, U. D. Grig, B. Gotsmann, W. HIberle, M. A. Lantz, H. E. Rothuizen, R. Stutz and G. K. Binnig, *IBM J. Res. Dev.* **44**, 323 (2000).
101. S. Jegadesan, S. Sindhu and S. Valiyaveettil, *Langmuir* **22**, 780 (2006).
102. N. Kumar, N. L. Abbott, E. Kim, H. A. Biebuyck and G. M. Whitesides, *Acc. Chem. Res.* **28**, 219 (1995).
103. L. Zhao and K. J. Adamiak, *Electrostat* **63**, 337 (2005).
104. M. Laan, J. Aarik, R. R. Josepson and V. Repan, *J. Phys. D* **36**, 2667 (2005).
105. S. F. Lyuksyutov, P. B. Paramonov, I. Dolog and R. M. Ralich, *Nanotechnology* **14**, 716 (2003).
106. S. Jegadesan, S. Sindhu and S. Valiyaveettil, *J. Nanosci. Nanotechnol.* **7**, 2172 (2007).
107. P. Taranekar, A. Baba, T. M. Fulghum and R. Advincula, *Macromolecules* **38**, 3679 (2005).

108. S. Jegadesan, P. Taranekar, S. Sindhu, C. A. Rigoberto and S. Valiyaveettil, *Langmuir* **22**, 3807 (2006).
109. S. Y. Jang, M. Marquez and G. A. Sotzing, *J. Am. Chem. Soc.* **126**, 9476 (2004).
110. B. W. Maynor, S. F. Filocamo and M. W. Grinstaff, *J. Liu, J. Am. Chem. Soc.* **124**, 522 (2002).

Jegadesan Subbiah received an MSc in Physics from Madurai Kamaraj University, India, and a PhD in Chemistry under Prof. Suresh Valiyaveettil at the National University of Singapore in 2007. Currently, he is working as a Postdoctoral Research Associate at the Department of Materials Science and Engineering, University of Florida, USA. His research interests include scanning probe lithography, nanofabrication of organic films, organic solar cells, polymer light-emitting devices and interface engineering of organic semiconductors. He has published over 24 peer-reviewed papers and made over 56 conference presentations.

Sajini Vadukumpully received an MSc in Applied Chemistry from Cochin University of Science and Technology, India. She is currently pursuing her PhD at the National University of Singapore, under Prof. Suresh Valiyaveettil. Her research interests include synthesis of carbon–rich nanomaterials and its applications, fabrication of organic thin films, and scanning probe microscopy.

Suresh Valiyaveettil completed his undergraduate and graduate education in India and received a PhD in Supramolecular Chemistry from the University of Victoria, Canada. He joined the National University of Singapore in 1998 and developed a multidisciplinary research program in materials chemistry. He has published more than 125 high-impact papers in peer-reviewed international journals in various fields, such as organic synthesis, polymer synthesis, nanomaterials, nanotoxicology and biomaterials, and given 40 invited talks at international conferences. Prof. Valiyaveettil is a member of national and international committees related to research and offers consulting in the above research areas. His research group is exploring multidisciplinary projects in organic/polymer synthesis, synthesis and characterization of biomaterials and nanomaterials/nanotechnology/nanotoxicity at the Department of Chemistry, NUS.

Index